NOTES ON THE EARLY HISTORY

OF THE

THE ROYAL REGIMENT OF ARTILLERY

BY THE LATE

COLONEL CLEAVELAND, R.A.

The Naval & Military Press Ltd

published in association with

FIREPOWER
The Royal Artillery Museum
Woolwich

Published by
The Naval & Military Press Ltd
Unit 10 Ridgewood Industrial Park,
Uckfield, East Sussex,
TN22 5QE England
Tel: +44 (0) 1825 749494
Fax: +44 (0) 1825 765701
www.naval-military-press.com

in association with

FIREPOWER
The Royal Artillery Museum, Woolwich
www.firepower.org.uk

In reprinting in facsimile from the original, any imperfections are inevitably reproduced and the quality may fall short of modern type and cartographic standards.

NOTES ON THE EARLY HISTORY

OF

THE ROYAL REGIMENT OF ARTILLERY.

BY THE LATE

COLONEL CLEAVELAND, R.A.*

In all wars before and after the conquest of England, it was the custom to use various sorts of instruments, engines, machines, spears, darts, javelins, bows and arrows, armour of iron for horsemen, footmen and charioteers, the care and provision of which in general, was committed to the Lieutenants of counties. The Master Bowyer, Master Fletcher, Master Carpenter, Master Smith, and other mechanics had patents and salaries from the Crown by way of retaining fees, and they were styled OFFICERS OF THE ORDNANCE.

The monks appear in these ancient times to have been the conductors of such military defences as were then made use of in earth or masonry, and intrusted with money for paying the expenses.

Upon a governor being appointed to a garrison, fort, or castle, a special commission was issued under the Great Seal appointing commissioners to take an inventory or "remain" of all the implements of war. The governor then entered into articles of agreement with the State, covenanting, that in consideration of the armour, ammunition, victual, provision, officers, soldiers, gunners, and artificers, according to an inventory annexed to the indenture, and in further consideration of his being supplied with money to pay the garrison, he covenanted to "defend and keep the same for his Highness and the State, against all traitors and enemies whatsoever :"—See two trials in the reign of Richard II., quoted by Selden in *Privilege of Baronage of England*, fol. 17, where John Whiston and the Lord of Gomeniz were condemned to death for delivering up their castles.

---

* The text of the original MS. is printed without alteration, but a few notes explanatory of ancient military terms have been contributed by Lieutenant-Colonel W. Lambert Yonge, R.A.

The first record of cannon used in the field is in the reign of Henry III.

The first record of the exact number of persons retained in the Ordnance, as well in time of peace as in war, appears in the reign of Edward III.

In the reign of Henry VIII., when the use of great ordnance became general all over Europe, England was famous for all sorts of workmen in the arts of manufacturing military weapons for the Tower of London. The Minories and adjacent places for a mile round were occupied as foundries, saltpetre houses, charcoal houses, sulphur houses, and shops for manufacturing all manner of warlike implements, fire arrows, fire darts, smoke balls, hand guns, harquebuses, which occasioned the employing a great number of mechanics in these abstruse branches: they took their abode chiefly in, and near the Tower, having patents and salaries from the crown as retaining fees.

Henry III. 1267. In the rebellion by the Duke of Gloucester, his majesty king Henry III. approached London with his army making daily assaults, when guns and other ordnance were shot into the city.—GRAFTON'S *Chronicles.*

Edw. II. 1322. His majesty Edward II. fought the battle of Leylade in Northumberland, where he lost his ordnance, which were conveyed into Scotland.—RASTELL'S, HOLINDSHEAD'S, and GRAFTON'S *Chronicles.*

Edw. III. 1327. In the first campaign of Edward III. against the Scotch, John Barbour the archdeacon of Aberdeen calls the English artillery "crakys of war." [1]

1333. His majesty marches an army into Scotland, and besieges Berwick, which capitulated; he also gained the battle of Holydown Hill.

1338. His majesty with his army and artillery crossed over to Flanders intending to attack France.

His majesty's ships, "Edward" and "Christopher," taken by the French; in this action they used guns and bowes and arbalister shot.—GRAFTON'S *Chronicles.*

1339. His majesty commences his march towards France, passing the town of Brussels, to the siege of Cambray; after besieging

---

[1] See his *Metrical Life of King Robert Bruce*, pp. 408, 9.

> Twa novelties that day they saw,
> That forouth in Scotland had been nane,
> Timbers for helmes was the ane
> That they brought then of great beautie,
> And also wonder for to see;
> The other crakys were of war,
> That they before heard never air.

it several days, his majesty from the lateness of the season, ordered it to be discontinued, and the army to immediately enter France.—Grafton.

1340. His majesty and army leave Ghent and besiege Tourney; after ten weeks assault, a truce was concluded, and the army with its great and small ordnance withdrawn to Ghent.

In the parliament held at Westminster, we find mention made of thirty-two tons of powder which was to have been furnished by one Thomas Brookhall.[1]

1342. The Scotch besiege the castle of Estrevelin with engines and cannon.—Grafton.

1343. His majesty with his army and artillery embark for France, and besiege Vannes.—Grafton.

1344. The number of persons of the ordnance retained by his majesty as well in time of peace as in war from the 18th to the 21st year of his reign, viz.:—

28 Masons.	24 Provisioners.
138 Carpenters.	60 Wargnores.
1 Cooper.	7 Armourers.
13 Smiths.	12 Artilleriens and gunners.
57 Engineers.	*Addl. MS.* 5758.

1346. His majesty embarks with an army and artillery for Normandy, capturing Harfleure, Cherburgh, &c., marched into the county of Evreux which he burnt, except the walled towns and castles; the king not assaulting them in order to spare his people and artillery. He afterwards passed the city of Bexways without any assault given, because he would not trouble his people nor waste his artillery.—Grafton.

26 Aug. His majesty fought the battle of Cressy. Mezeray says that king Edward struck terror into the French army with five or six pieces of cannon, it being the first time they had seen such thundering machines.—Rapin's *History of England.*

1347. His majesty besieges Calais: he determined not to attack the town in order that he might spare his people and artillery, but

---

1 The following is from Grose's *Military Antiquities*, Vol. I. p. 378.—"In the work called Cotton's *Abridgment of the Records of the Tower of London*, p. 24, there is a strange mistake respecting gunpowder; it being there said, that a pardon was directed to be made out 14 Edward III. to Thomas de Brookhall, for a debt of 32 tons of powder, and in the index it is added (by way of note) "before its pretended invention." The original in the rolls of Parliament stands thus:—'Item, pur Thomas de Brookhall, pur trent et deux toneux *de pomadre*, des quex il est chargé sur son acounte du temps qu'il estoit assigné de faire divers purveances a lœps le roi en la countée de Kent.' *Pomadre* is cider, instead of gunpowder, and probably provided for the king's drinking."

famish them by a long siege. On the approach of the French king he fortified a castle between the town and the sea, "with springaldes,[1] bombardes, bowes, and other artillery;" he also secured his lines, by causing his navy to lie near the shore, every ship well filled with bombardes and other artillery.[2]

The following persons at the siege of Calais were paid by the ordnance,—masons, carpenters, smiths, engineers, tent-makers, miners, gunners armed and those that had the care of the artillery, numbering 314; some at one shilling, others at tenpence, sixpence, and threepence per diem.—RAPIN and GRAFTON.

Rich. II. 1384. To John Walsh receiver of the king's provisions at Cherburgh, in money paid to him by John D'Arundell, marshal of England, late keeper of the castle and town of Cherburgh at the time he was discharged from the custody aforesaid for the undermentioned things, remaining there for the king's use for the defence and provision of the castle and town aforesaid, viz.—

For ten guns to throw stones, two of which are of iron and eight of other metal; seven of the said ten guns casting large stones 24 inches in circumference, and the three remaining casting large stones 15 inches in circumference; 200 lbs. of powder, 26 lbs. of saltpetre, and 24 lbs. of pure sulphur, were paid for at the exchequer in the account of the 17th day of June in the eighth year of our reign, &c.—Issues of the Exchequer by FREDERICK DEVON. See *Athenæum*, No. 503.

1386. Two French ships taken and brought into Sandwich; on board was found a master-gunner, who led the English at Calais, also divers great guns and engines to beat down walls, also a great quantity of powder.—HOLINDSHEAD.

1398. His majesty with a great power and artillery embarked for Ireland and brought the greater part of that realm into good order.—GRAFTON.

Henry IV. 1400. His majesty with a great provision of men, munition, and artillery march into Wales, and drive the French out of that country.—GRAFTON.

1405. His majesty marched with his army into the north, and with his artillery besieged Berwick: the first shot overthrew part of one of the towers which caused them to surrender. The army afterwards proceeded to Wales.—HOLINDSHEAD.

---

[1] *Springal.* An ancient military weapon for casting stones and arrows:—

"And sum that wente to the wal
With bowes and with springal."—*Bevis of Hamtoun*, p. 159.

"Trybget,* springlas, and also engyne
They wrougt owre men full meckyl payne."—*Archeologia* xxi. 61.

* Trabucci.

[2] "Siege" was thus merely a "*sitting down* before a town and investing it."

Henry V. 1414. Warrant from his majesty to Nicholas Merburg, master of the ordnance, and to John Louth his clerk, assigning the necessary workmen, and materials, with authority for pressing all sorts of carriages required for that service.—RYMER'S *Fœdera.*

1415. His majesty embarked at Southampton with his army and artillery, with boats covered with leather for crossing rivers, and landed at Havre de Grace.

He invested Harfleure, where his artillery was plied with such success that a considerable breach was made, which caused the enemy to capitulate; his majesty after repairing the bulwarks, walls, rampiers, and furnishing it with provisions and artillery, moved towards Pouthoise,

The following persons of the ordnance were present at the siege, viz.—

John Greyndon, Knight, with miners ...		120
Gerald Van Willighen, Haynes Joye, Walter Stotmaker, Drovankesall Coykyer	master gunners with others	25
Each with two servitour gunners[2]		50
Nicholas Brampton, stuffer of bacynets[1]		
Armourers		12
Carpenters		124
Wheelers		6
John Benet, mason, with other labourers		120

25 Oct. His majesty fought the battle of Agincourt, wherein the above artillery were engaged.—NICOLAS'S *Agincourt,* 2nd Edit.

1417. A writ directed to John Louth, clerk of the ordnance, and John Benet of Maidstone, mason, reciting that a sufficient number of masons and labourers had been assigned for making seven thousand stone shot for guns of different sorts, with a sufficiency of stone for the same, as well in the quarries of Maidstone or elsewhere as should be most for the benefit of the service.

The workmen to be kept till the whole are complete, and men to be impressed for carting, boating, or other carriage of the said stones.—RYMER'S *Fœdera.*

July. His majesty with his army, artillery, and a thousand masons, carpenters, and other labourers, sailed from England and landed at Tongue in Normandy (retaining a few vessels for transporting his artillery) which he besieged and became master of, 9th August.

9 Sept. Caen surrendered the 9th September, also several other towns and castles in this year.—HOLINDSHEAD and RAPIN.

His majesty who had commenced his campaign in August of this year, and which lasted the whole winter, without interruption or

1 *Bacinets* were light helmets, so called from their resemblance to a bason, and were generally without vizors. Bacinets (or bassinets) were worn in the reigns of Edward II. and III. and Richard II. by most of the English infantry, as may be repeatedly seen in the Rolls of Parliament and other public records.—GROSE.

2 Afterwards called "mattrosses."

respite, finished with besieging the town and castle of Falaise the 1st December.

1418. 16 Feb. Falaise surrenders.

March. April. In the months of March and April he possessed himself of St Lo, Carentan, St Sauveur le Velcomte, and many other towns in Normandy.

May. Evreux surrenders.

August. Cherbourg surrenders after a siege of three months.

Sept. Rouen the last remaining fortress in Normandy belonging to the King of France, besieged, and capitulated after five months.

1419. 19th Jan. Rouen surrenders.

Normandy brought into the king's subjection.

Henry VI. 1420 to 1425. The army continued in Normandy, occupying the several fortresses, and capturing others, and the artillery were conspicuous in firing great stones out of great guns against their walls, his majesty fortifying the same with men, ordnance and artillery.—HOLINDSHEAD and GRAFTON.

The English defending the town of St James de Leitson on the frontiers of Normandy, issued out attacking the besiegers, and capturing 14 great guns and 40 barrels of powder.—GRAFTON.

1428. The earl of Salisbury besieges Orleans, which he encompassed with forts to the number of sixty, of these six were considerably stronger than the rest commanding the principal avenues to the city, and in which he placed his cannon.

When the besieged perceived that they were environed with fortresses and ordnance, they laid gun against gun, and fortified towers against bulwarks, and within made new rampiers, and built mud walls to avoid cracks and breaches, which might by violent shot suddenly ensue.—GRAFTON and RAPIN.

1433. The earl of Arundel wounded by a shot from a culverin in his attack of the castle of Gerborye, of which hurt he shortly died.—GRAFTON.

Item for his (Duke of Somerset's) shipping and his ordnance.

"Be there made a privy seal to Gilbert Parr, master of the king's "ordnance, to deliver to John Downson, master of the ordnance of "my Lord of Somerset, 4000 lbs. of saltpetre, 3000 sulphur, 3000 "bowes, 3000 sheaves of arrows, 200 gross of strings, &c."—NICOLAS'S *Privy Council Proceedings*, Vol. V.

1449. The Duke of Somerset, the king's lieutenant in France, acquaints the king and council, that he required both ordnance and artillery for the defence of Normandy in case of its being attacked.

1455. Thomas Vaughan, Esq., the then master of the ordnance, presented a petition to His Majesty stating, "That for as muche as ther is noow housing certaynly assigned for your ordenaunce to be kept, for lak whereof ther hath growne grete hurt and dayly doth unto the said ordenaunce and other stuffe longing to his said Office." And in consideration thereof he prayed His Majesty to grant for its use "All the grounde and soille called ye Tour Wharf, from the Watergate of the Tower now called the Traitor's Gate, unto the Gate of St Catherine's, together with all maner of howsing, and other appurtenaunces sette uppon the same," which was accordingly granted under the royal signature.—BAYLEY's *History of the Tower of London.*

1456. Commission to John Judd as master-general of the ordnance. *Excerpta Historica*, p. 10.

Edw. IV. 1471. King Edward IV. landing at Ravenspur in Yorkshire, brought over and introduced hand-guns or muskets.[1] In his battles with the Earl of Warwick, both armies were furnished with great artillery.—HOLINDSHEAD.

1474. 4 July. The king embarks for France with a large army and artillery, with artificers and labourers.—HOLINDSHEAD.

1482. The king collected his army and prepared his artillery to invade Scotland, a thousand men of the army were appointed to give their attendance on the ordnance. The whole embark under the Duke of Glocester, and arrive at Berwick, which they besiege and capture the 24th Aug. The Duke placed a garrison of Englishmen, and artillery sufficient to defend it against all Scotland for six months.—HOLINDSHEAD.

1483. Feb. 26. A warrant to the constable of the Tower to deliver to Roger Beckley eight serpentines upon carts, 28 hackbushes with their frames, one barrel of touch powder, two barrels of serpentine powder, &c.

Edw. V. 16 June. A warrant for delivering unto Thomas Tunstall, two serpentines, two guns to lie upon walls, twelve hackbushes, two barrels of gunpowder, &c.

Rich. III. Raufe Bygood was appointed to the mastership of the ordnance during life with a hundred marks fee for himself, and the wages of sixpence per diem for a clerk, and sixpence for a yeoman. John Stoke the office of clerk of the ordnance within England or elsewhere for the time of his life, with wages of sixpence per diem. Richard Warrington the office of the artillery within the town of Calais with wages of one shilling per diem, and sixpence for a yeoman under him

[1] Note this definition of "muskets."

for life. To Patrick de le Mayte, gunner, the office of master-gunner within the Tower of London, to Henry Grey the younger, squire, a confirmation of the office of the keeping of the armoury within the Tower of London. To Theobald Ferrount, gunner, an annuity of £12. To William Tempell the office of yeoman of the ordnance.—*Harl. MS.* No. 433, pp. 31, 52, 105, 157, 178.

Hen. VII. 1485. Sir Richard Gilsford appointed master of the ordnance.—GROSE'S *Military Antiquities.*

1491. Sir Edward Poinings with bold soldiers and sufficient artillery, embark for Flanders and besiege Sluis, which capitulated.

6 Oct. The king embarks with his army and artillery for Calais and besieges Boulogne. After breaking the walls and sore defacing them with his battering pieces, peace was concluded between them.—HOLINDSHEAD.

1496. His majesty marches with a populous army, and plenty of artillery, out of the city for the attack of the Cornish rebels on Blackheath whom he defeated.—HOLINDSHEAD.

1497. The Earl of Surrey with an army and artillery march into Scotland in pursuit of the Scotch army; he besieges Haiton Castle which surrendered in a few hours. The Earl caused his miners to raise and overthrow the fortress to the plain ground.—HOLINDSHEAD.

1498. In this year all the gardens which had been continued out of mind, without Moorgate of London, were destroyed, and of them was made a plain field for archers to shoot in.—HOLINDSHEAD.

Hen. VIII. James king of Scotland about this time gave liberal possessions to Robert Borthwick, a notable artificer for making of field pieces and other guns; for which liberalities he should make certain great pieces in his castle of Edinburgh, whereof there are many yet to be seen in Scotland with this superscription,

"Machena sum Scoto Borthwick faboricata Roberto."

HOLINDSHEAD, Vol. V. p. 470.

1511. Sir Thomas Darcy with 1500 archers and others on an expedition to Spain.

Sir Edward Poinings, knight, with 1500 archers and artillery, commanded by Thomas Hart, governor or chief gunner, embarked on an expedition to Flanders, to assist the Lady Margaret, Duchess of Savoy. They besieged and won the castle of Brimnoist, and took possession of the town of Aiske.—HOLINDSHEAD.

1512. Thomas Grey, Marquis of Dorset, embarked with an army and artillery for Spain; they landed at Passage on the coast of Briskey; Sir John Stiles ordered 200 mules to be purchased for the artillery, and prepared for attacking the French in Gascoyne.—HOLINDSHEAD.

1513. Sir Sampson Norton appointed master of the ordnance.

21 July. King Henry with an army, and the master of the ordnance, with the king's artillery,[1] as fauchons, slings, bombardes, carts with powder, stones, bowes, arrows, &c.; the whole number of carriages 1300, the leaders and drivers of the same 1900 men, marched from Calais for the siege of Terouenne. On the march, by the negligence of the carters, a great curtall[2] (called the John Evangelist) was overthrown into a pond of water, also a bombarde of iron (called the Red Gun) overthrown in a lane. The first gun was captured by the French and carried into Boulogne, the other was recovered by the Lord Barnes, captain of the pioneers and labourers, protected by the Earl of Essex and his company.

On the 4th August his majesty arrived before Terouenne, enclosing his camp with artillery, as fauchons, serpentines, cart hackbushes, and tried harrows, &c., and with his great ordnance did sore beat the walls; and Sir Alexander Buinam, a captain of miners, caused a mine to be enterprised to enter into the town.

16 Aug. Battle of Spurs. On the 16th the master of the ordnance threw five bridges across the river, over which the army and great ordnance passed, and attacked with success the French army, when the English artillery (culverins) came into play. The batteries having breached the walls in several places, Terouenne surrendered the 18th. The walls and fortifications were razed,

18 Aug. the town burned and the ordnance sent to Aire for the king's use.

23 Aug. His majesty marched to invest Tournay, the siege commenced the 23rd with 21 pieces of great artillery, and surrendered the 29th.

1513. Sept. The Earl of Surrey collects his forces at Alnwicke, Sir Nicolas Appleyard, master of the ordnance; they march towards the Scotch army, and on the 9th gained the battle of Flodden, wherein King James was killed. The next day a Scotch force appearing in the field, William Blackenall, who had the chief rule of the British ordnance, caused such a peal to be shot off at them that the Scots fled. Two and twenty pieces of ordnance were taken, amongst them seven

1 Note the detail of this "artillery."

2 The piece of ordnance called a "curtall" is seldom mentioned by the early writers on arms and armour, and it is therefore difficult to ascertain the nature of the gun, though from the similarity of the word to that used to signify "a horse or a dog whose tail has been very closely cut," it may be inferred that the piece of ordnance was one of those which like the *bastard* and *minion* differed in dimensions from the "ordinary" gun of the period. The "curtall" is mentioned by Hall in his Chronicle of Henry VIII.

culverins of large assize, called by King James, the "Seven Sisters."[1] —HOLINDSHEAD.

The master gunner of the English part opened his fire, slew the master gunner of Scotland, and beat all his men from their ordnance, so that the Scotch ordnance did no harm to the English, but the English artillery shot into the midst of the king's battle, and slew many persons.[2]

1516. His majesty ordered 1200 masons, and carpenters, and 300 labourers to the city of Tournay.

1519. Sir Edward Gilford, master of the armoury.—HOLINDSHEAD.

1521. Great brass cannon and culverins first cast in England by John Owen.—STOWE.[3]

1523. 6 July. The Earl of Surrey with an army, and artillery commanded by the master of the king's ordnance, Sir William Skevington, having landed the Emperor in Biscay, returned with his fleet and made a descent on the coast of France, near Morlies, marching thither and assaulting the town won it, for the master gunner, Christopher Morris, having there certain falcons, with the shot of one of them struck the lock of the wicket in the gate, so that it flew open, and then the same Christopher Morris and other gentlemen with their soldiers in the smoke of the guns pressed to the gates, and finding the wicket open entered, and so finally was the town of Morlies won and put to sack. The army shortly after returned to England.—HOLINDSHEAD.

23 Aug. Admiral Sir William Fitzwilliam with his fleet, soldiers, and field artillery, under the master gunner, Christopher Morris, sail for the French coast, land near Trayport, after severe fighting they re-embark after burning seven fine ships, destroying their bulwarks, and capturing twenty-seven pieces of ordnance.

---

1 "They saw, slow rolling on the plain
Full many a baggage cart and wain,
And dire artillery's clumsy car
By sluggish oxen tugged to war.
And there were Borthwick's sisters seven,
And culverins which France had given.
Ill omen'd gift! the guns remain
The conqueror's spoil on Flodden's plain."
Marmion, Canto IV. Division 27.

2 The following extract will give an idea of the value of pieces of artillery at this period:—"The king* went in person against it,† and for its reduction borrowed from the castle of Dunbar, then belonging to the Duke of Albany, two great cannons, whose names were 'Thracon mouthed Mow and her Marrow,' also 'two great botcards and two moyans, two double-falcons, and four quarter-falcons,' for the safe guiding and re-delivery of which, three lords were laid in pawn at Dunbar."—*Marmion*, Note 15 to Canto 5.

* James V. † Tantallon Castle.

3 Formerly apparently procured from France, "which France had given,"

1524. His majesty directed an army under the Duke of Suffolk with two thousand six hundred labourers and pioneers to be transported to Calais. The first enterprise was the winning the castle, called Bell Castle, by the Lord Sands and the Lord Ferrers, who did so much by the power of battery that they surrendered, and Sir William Skevington was directed to raze it down to the ground.

26 Oct. They afterwards besieged Bray, which was carried by assault the 20th October, after five hours battering with their ordnance.

27 Oct. Sir William Skevington opened his batteries against Mondedier with such force that after four hours battering the walls were overthrown and made assaultable, when the garrison surrendered.

13 Nov. The Duke assaulted and carried Boghan Castle, wherein were found 76 pieces of artillery, as bombardes, curtaux, and demi-curtaux, slings, cannon, volgers, and other ordnance. The army returned to Calais leaving his great ordnance at Valencenes.—HOLINDSHEAD.

1526. Peace concluded with the French.

1530. The Irish spoil and burn the county of Kildare; his majesty sent the Earl of Kildare to Ireland with Sir William Skevington, master of the king's ordnance, and divers gunners with him, which so politically ordered themselves that the enemies were glad to offer amend, and Sir William Skevington the next year returned to England, leaving the Earl of Kildare for the king's deputy.—HOLINDSHEAD.

1532. The master of the king's ordnance lodged in the "Bryck Tower," of the Tower of London, also the Ordnance Office in the same.—BAYLEY'S *History of the Tower*.

1537. To the many laws, ordinances, and regulations issued for the support of archery, may be added the institution of the Artillery Company, which was incorporated by the Patent of Henry VIII. in the year 1537, to Sir Christopher Morris, knight, master of the ordnance, Anthony Knevet and Peter Mewtes, gentlemen of the Privy Chamber, overseers of the fraternity or guild of St George, granting licence to them to be overseers of the science of artillery, viz. for long bows, cross bows, and hand guns; and the said Sir Christopher Morris, Cornelys Johnson, Anthony Anthony, and Henry Johnson, to be masters and rulers of the said science of artillery during their lives, &c.

Sir Christopher Morris appointed master of the ordnance.—GROSE'S *Military Antiquities*.

1543. The first pieces of *iron*, both cannon and mortars, ever made in England, were manufactured at Buckstead in Sussex by Peter Baud, Ralph Hoge, and Peter Van Colin.—STOWE.

1544. 4 May. The Earl of Hertford with an army (and artillery commanded by Sir Christopher Morris, master of the ordnance), embarked for Scotland, landing near Leith, which town they immediately captured.

On the 6th the army marched towards Edinburgh, when Sir Christopher Morris with his gunners and ordnance, beat the Scots from their ordnance, and opened the Canongate with the shot of their great ordnance.—HOLINDSHEAD.

His majesty's warrant to Sir Thomas Seymour, master of the ordnance to provide ordnance, carriages, ammunition, and habiliments of war, with officers, their servants, conductors, gunners, wagoners, pioneers, carpenters, bowyers, fletchers, and other artificers, with the freight to accompany the army in France.

	per diem. £	s.	d.
Sir Thomas Seymour, master of the ordnance	1	6	8
do. conduct money at 4d. per mile			
do. coat money for 20 servants at 4d. each.			
A horse tent.			
Sir Thomas Darcie, master of the armoury	1	0	0
Sir Christopher Morris, knight, lieut. of the ordnance	0	10	0
A clerk for him, Robert Morgan	0	2	0
6 servants, each	0	0	6
Burnardyne de Vallowayes / John Bassett — master gunners each		4	0
209 gunners, each	0	0	7
157 artificers, each	0	0	8
John Cookson, chief conductor of the train	0	6	0
A clerk to him	0	2	0

	per diem. £	s.	d.
John Verney, overseer of the king's great mares for the train of artillery	0	4	0
William Heywood, assistant to him	0	1	0
Thomas Mulberry / Harry Hughes ... — guides of the said mares, each	1	0	0
6 conductors of the ordnance.			
20 carters, each	0	0	6
William Reyherne, captain of the pioneers	0	4	0
100 pioneers, each	0	0	8
John Rogers of the privy ordnance and weapons.			
15 gunners appointed to the brass pieces about the king's tent.			
55 gunners appointed to the shrympes with two cases each.			
4 carpenters.			
4 wheelers.			
3 armourers.			

Charles Walman, an officer employed to choose the gunpowder.

The pioneers received two shillings apiece transport money from Boulogne to Dover, and conduct money from Dover to their dwelling places; 4*d.* a mile for the captain, and ½*d.* for every pioneer.—*Harl. MS.* 5753.

14 July. His majesty with an army passed over to the siege of Boulogne, where Sir Christopher Morris, master of the ordnance, with his great artillery and mortar pieces so battered the walls from three several places, that there were very few houses left; the breaches being practicable, the assault was made by the Lord Admiral Dudley, under the protection of the artillery, which kept up a continual fire on the breach during their advance. After a severe conflict the assailants were recalled, shortly after which the governor capitulated.

Sir Christopher Morris was hurt by a hand-gun at the siege.—HOLINDSHEAD and RYMER.

1545. 14 Feb. The Earl of Hertford informs the Lord Cobham that the ordnance is arrived, but the stock of the demi-cannon is broken, and that therefore he sends Wenlock, the master-gunner of Boulogne, to choose such another as he shall think meet.—*Harl. MS.* 284.

20 July. His majesty's fleet sailed out of Portsmouth harbour to attack the French at St Helens. (One of the fleet the "Mary Rose" was drowned in the midst of the haven by reason she was overladen with ordnance, and had the ports left open, which were very low, and the great artillery unbreeched, so that when the ship should turn, the water entered and suddenly she sank. In her was Sir George Carew, knight, and 400 soldiers under his guiding; only 40 escaped. Several of the guns were raised in the year 1835-6 and brought to the Arsenal at Woolwich).

1546. The Earl of Surrey writes over to the king's council, that the French had cast much larger cannon than had yet been seen, with which they imagined they should demolish Boulogne.

The Earl of Hertford with an army and artillery, Sir Thomas Wiat, master of the ordnance, were sent over to Calais.—HOLINDSHEAD.

Edw. VI. 1547. Allowances to officers and ministers of the ordnance in Edward VI. reign.

per annum.	£	s.	d.
The master of the ordnance, fee	151	6	8
The lieutenant	100	10	0
Surveyor	36	10	0
Clerk	38	8	0
Yeoman	27	7	6
per diem.			
Master-gunner	0	1	0
Gun stone maker	0	0	6

per diem.		£	s.	d.
Saltpetre maker		0	0	6
Gunfounderer		0	1	4
Gunsmith		0	0	6
Artificers, each		0	0	4
10 master overseers, each		0	0	8
58 gunners	20, each at	0	1	0
	12 "	0	0	8
	24 "	0	0	6
	2 "	0	0	4

Besides gunners in the several forts and castles in England.—*Harl. MS.* 240.

Sir Francis Flemynge, knight, appointed master of the ordnance.—HOLINDSHEAD and GROSE.

10 Sept. The Duke of Somerset marched into Scotland, gained a complete victory at Musselburgh, taking eighty pieces of cannon in his short expedition. Sir Francis Flemynge, master of the ordnance, with 15 pieces of great ordnance, and other pieces; John Man, treasurer of the ordnance, Sir Richard Leigh, knight, deviser of the fortifications,—Cornelius comptroller of the ordnance, John Brenne with 1400 pioneers accompanied the artillery with the army.—HOLINDSHEAD.

1548. Sir Phillip Hoby, appointed master of the ordnance.

The English artillery in the reign of Edward VI., and about the year 1548, consisted of

		£	s.	d.	£	s.	d.
1 master of artillery...	Sir John Hoby......(fee)	0	0	0	151	11	8
1 lieutenant......... ..	Sir Francis Fleming „	0	0	0	66	13	4
1 surveyor	Anthony Anthony... „	0	0	0	36	10	4
1 clerk	John Rogers .... „	0	0	0	12	3	4
	In room of servants..............	0	0	0	12	3	4
1 yeoman	Thomas Sheventon... „	0	0	0	9	2	6
	In room of servants..............	0	0	0	18	5	0
1 master-gunner......	Christopher Gold ... „	0	1	0	18	5	0
1 gun stock maker...	Symond Fernes ...... „	0	0	0	9	2	6
2 gun founders	John Owen .... „	0	0	8	12	3	4
	Thomas Ogey......... „	0	0	0	12	3	4
1 gunsmith............	John Anthony........ „	0	0	6	9	2	6
1 artificer or engineer	John Pudney .... „	0	0	4	6	1	8
1 master carpenter...	John Johnson......... „	0	0	8	12	3	4
109 gunners whereof	15 at 12d. a day each...........	0	15	0	273	15	0
	12 at 8d. „ „	0	8	0	146	0	0
	80 at 6d. „ „	2	0	0	730	0	0
	2 at 4d. „ „	0	0	8	12	3	4
	Total charge of the Artillery for one year				£1547	9	2

In the 109 gunners above mentioned are included fees to the following persons, viz. :—

To the clerk John Rogers......at 8d. a day.	To the yeoman ....at 12d. a day.
„ master-gunner.........at 12d. „	To John Owen, gun founder, at 12d. „

So that 105 effective gunners remain.

These two last accounts are taken from a manuscript of the Rev. W. Gostling's and very obligingly communicated by Cap. W. Gostling of the Royal Artillery.

15 Jan. Sir Philip Hoby writes to the Duke of Somerset from Brussels, desiring that due care may be taken of the office of the ordnance, which was incorporated by king Henry VIII., and not since confirmed which, with the non-payment of their debts, has brought them into decay.—*Harl. MS.* 623.

An order in council to fortify Lowder in Scotland. Sir Thomas Palmer and Sir Thomas Holcroft were appointed to go thither with a convenient number of men of war and pioneers to see that town fenced with trenches, rampiers, and bulwarks.—HOLINDSHEAD.

1549. 4 Oct. The Lord Protector and Council, to the Lord Cobham, Deputy of Calais, desiring him to send speedily 20 of the ablest gunners, which were at Newhaven, and now remain in Calais, by sea to the Lord Clinton at Boulogne.—*Harl. MS.* 284.

The Earl of Warwick, with an army and artillery, is ordered against the Norfolk rebels, whom he overcomes after hard fighting.—HOLINDSHEAD.

The Earl of Rutland ordered to raise the fortifications of Hadington, and remove the ordnance to Berwick.—HOLINDSHEAD.

1552 30 April. Through the negligence of the gunpowder makers, a certain house near the Tower of London, with three carts of powder was blown up and burnt with 15 gunpowder makers.—HOLINDSHEAD.

Queen Mary 1553. Sir Edward Brair appointed master of the ordnance.

Sir Edward Brair with certain field pieces encountered the rebels at Charing-cross.—GROSE AND HOLINDSHEAD.

1555. Sir Richard Southwell, master of the ordnance and armoury.—HOLINDSHEAD.

1557. An army with artillery embarked under the Earl of Pembroke for the Netherlands, and joined the Spanish army, commanded by the Duke of Savoy. The Lord Robert Dudley, Earl of Leicester, master of the ordnance, and Sir Richard Leigh, trench master, Edward Chamberlain, captain of the pioneers.—HOLINDSHEAD.

Establishment of the artillery for the above expedition, who with the Lord Robert Dudley were present at the siege and capture of St Quintins.

	per diem. £	s.	d.		per diem. £	s.	d.
The master of the ordnance	1	6	8	3 smiths	0	3	0
His lieutenant	0	13	4	3 guides of the ordnance	0	4	0
Master of the carriages	0	10	0	12 carriages	3	0	0
The trench master	0	5	0	A drum	0	1	0
A chaplain	0	1	0	A fife	0	1	0
A clerk of the ordnance	0	2	0	120 smiths	1	5	0
2 clerks	0	2	0	10 halberdyers	0	10	0
A surgeon	0	1	0	Harquebutters on horseback for the lieutenant	0	6	4
6 boweyers	0	6	0				
6 fletchers	0	6	0	Master-gunner	0	3	0
3 carpenters	0	3	0	12 gunners	0	16	0

Harl. MS. 6844.

1558. John Hiefield, master of the ordnance, was sent from Calais to give due advertisement of the French king's intention of attacking Calais, and to have a supply of things necessary for mounting of the great artillery, whereof he had charge; he was made prisoner on its capitulation. Master-gunner Horslie in the defence of Newnam Bridge, lost his head by the first shot from the French batteries.—HOLINDSHEAD.

Sir William Pelham appointed master of the ordnance.—GROSE'S *Military Antiquities*.

Queen Elizabeth 1559. Ambrose Dudley, Earl of Warwick, appointed master of the ordnance.

Mr. Gower, master of the ordnance in the northern parts of England under the Lord Grey of Wilton.

			£	s.	d.
John Owrde, gentleman, master of the ordnance at Berwick, with pay, per annum...			20	0	0
40 gunners paid.	10 gunners	per annum, each	9	2	6
	22 do.		9	0	0
	8 do.		6	8	4

From *State Papers* of Sir Ralph Sadler.

3 Oct. Sir Ralph Sadler, and Sir James Croft to the Lords of the Privy Council.

"It may like your good lordships to understand, that whereas our late Sovereign Lady, Queen Mary, in her time addressed her letters to the Earl of Westmoreland, then her highness lieutenant here, that he should take order for the allowance of two pounds of corn powder every month, to be given to every harquebutier in these bands, of her gracious liberality and reward, to the intent they might by the use and exercise of their pieces learn to be more perfect in that feat. Whereupon the said Lord Lieutenant wrote his letters to Mr. Gower, then master of the ordnance in these north parts, willing him to make the said allowance by the warrant of Sir John Brende, then muster-master here, from time to time, which was performed accordingly untill the 1st of May last past; since which time the said Gower hath no further warrant; notwithstanding that, ever since unto this present the soldiers have had their powder as they had before of the Queen's Majesty Store, thinking to enjoy still their foresaid allowance. There is now a question risen thereof betwixt the said Gower and the captains and soldiers here of the said bands, for that the said Gower requireth to stop so much money in the treasurer's hands of their pay as would satisfy for the powder which they have had of him since the said 1st of May; and on the other side the said captains and soldiers do allege, that they knew none other but that they should still enjoy their said allowance, because they had no sufficient warning to the contrary, or else the poor soldiers would not have spent so much powder as they have done, which if they should now be compelled to pay for, should be to their great hinderance.

"The said Gower sayeth he gave them warning thereof, and they deny it, so that hereof ariseth the controversey. Wherefore we are become humble suitors to your Lordships that it may please you to have some consideration of the same; and though we think that it had been much better to have relieved the soldiers some other way, because we like not the president, yet since it is past, and that thereby the harquibutiers here are become so perfect in their feate, that for so many we think there be no better of no nation, We humbly beseach your good lordships, that it may please you to direct your letters unto the said Gower, for his better warrant to make the said allowance till this present, and from hence fourth it may cease, whereby though the Queen's Majesty be the loser, yet the controversey would be ended, and the captains and soldiers here well contented."

16 Dec. Mr. Secretary Cecill to Sir Ralph Sadler and Sir James Crofts.

"It is wished we should take possession of the Fryth for our greater succor, and that Lord Grey should do this exploit, and have, besides the footmen, 600 horsemen with lances and pistolets, and 14 or 1500 light horsemen, and five or six good brass pieces for battery."

22 Dec. Sir Ralph Sadler recommended to Mr. Secretary Cecill that a large proportion or ordnance should be sent with Lord Grey.

1560. Thomas Gower, master of the English ordnance present with the army in the attack of the French at Leith.

Pay of the gunners of the great ordnance at Berwick.

	per diem. s.	d.		per diem. s.	d.
1 master-gunner at	4	0	4 quarter-masters, each	1	4
His mate „	2	0	44 gunners „	1	0

At Carlisle.

10 gunners at per diem, each 8*d.*

At Worke Castle.

	s.	d.		s.	d.
1 master gunner at per diem	1	0	His mate at per diem	0	10

2 gunners, at per diem each 9*d.*

In the North.

Thomas Gower, master of the ordnance	6	8	John Flemynge, master-gunner	3	4
			His mate	1	8
John Bennet, master of the ordnance, his clerk and his men	12	4	4 quarter-masters, each	1	0
			44 gunners, each	0	10

William Bromfield, lieutenant of the ordnance.—*Harl. MS.* 5752.

1561. Queen Elizabeth ordered Berwick to be well fortified, and directed many brass cannon to be cast, and encouraged the making gunpowder.

The Earl of Warwick lands with his army at Newhaven in Normandy.

1562. William Bromfield sworn master of the ordnance (under the Earl of Warwick, captain-general of the English forces in Normandy), killed in the defence of Newhaven in 1563.

Sir John Portinarie sent from England as chief engineer under the Earl.

The Earl of Warwick wounded.

1565. Robert Thomas, master-gunner of England, killed by a piece of one of the chambers, when attending the firing a salute at the marriage of the Earl of Warwick, general of the ordnance.

1566. Edward Randall, Esq., lieutenant of the ordnance sent into Ireland, where he lost his life in an engagement with the Irish.—HOLINDSHEAD.

1569. Humfry Barwicke, master of the ordnance at Berwick.
Accounts rendered by Sir Ralph Sadler, treasurer to her majesty touching the great ordnance, viz.—

	per diem. £	s.	d.
Humfry Barwicke, master of the ordnance for the field, for his own entertainment at	0	10	0
3 light horsemen attending upon the ordnance, each	0	1	4
3 footmen attending upon carriages, do	0	0	8
1 clerk	0	2	0
1 master-gunner	0	2	0
1 mate	0	1	0
2 gunners....................................	0	0	8
1 master-carpenter and his man......	0	1	8
1 master-smith, 1 wheeler, each......	0	1	0
2 men .... „	0	0	8
4 carpenters................. „	0	0	10
8 do. .... „	0	0	8

	per diem. £	s.	d.
Humfry Barwicke, capt. of pioneers	0	4	0
100 pioneers, each	0	0	6
Humfry Barwicke, for sundry charges touching the said great ordnance, as by a book of the parcels doth appear, paid by warrant of the said lord lieutenant, dated 29 January, 1569, (70) annexed to the said book together with his acquit.			
The same Humfry Barwicke, by warrant, dated 3rd December, 1569, for conveyance by water of armour, and ammunition from Kingston-upon-Hull to York, viz., to the master of the "Keale" 72s., and the marrinors for their charges	4	9	4

John Wildinge, master boweyer of ordnance in the north parts, for the charges of bringing by ship of certain armour and ammunition from Newcastle unto Kingston-upon-Hull, and returning back by warrant, dated 3rd December, 1569.

1570. The Earl of Sussex, lord lieutenant, with an army and artillery, both field and battering train, marched from Berwick, into Scotland, destroying several towns and castles.

1572. 16 Feb. A warrant from her majesty to the Earl of Warwick, master of the ordnance, to send from the Tower 50 carts of cornpowder and 5 carts of serpentine powder to Thomas Sutton, master of the ordnance at Berwick.—*Addl. MS.* 5754.

Mar. 25. An order in council for choosing gunners out of the most likely and active persons belonging to the several trades in London, who were trained up in warlike feats, mustered three times a week, sometimes in the artillery yard, teaching them to handle their pieces, sometimes at the Mile's End, and in St. George's Field.—HOLINDSHEAD.

1573. 28 July. A warrant from her majesty to the Earl of Warwick, master of the ordnance, to send powder and stores to Jacques Wingfield, Esq., master of the ordnance in Ireland.

1575-1577. Similar warrants to the above.—*Addl. MS.* 5754.

1574. The names of our great ordnance are commonly these.

NAMES.	Weight.	Diameter.	Weight of shot.	Scores of carriage.	Charge of powder.	Height of bullet.
	lbs.	inches.	lbs.		lbs.	inches.
1 Robinet	200	1¼	1	...	½	1
2 Falconet	500	2	2	14	2	1¼
3 Falcon	800	2½	2½	16	2½	2¼
4 Minion	1100	3¼	4½	17	4½	3
5 Sacre	1500	3½	5	18	5	3¼
6 Demi-Culverin	3000	4½	9	20	9	4
7 Culverin	4000	5½	18	25	18	5¼
8 Demi-Cannon	6000	6½	30	38	28	6¼
9 Cannon	7000	8	60	20	44	7¾
10 E Cannon	8000	7	42	20	20	6¾
11 Basiliske	9000	8¾	60	21	60	8¼

1578. From Grose's *Military Antiquities*, Vol. I. p. 198, was taken the duties of the MASTER OF THE ORDNANCE in the field, as obtained from an ancient manuscript of this date:—

"First, it is the office of the Master of the Ordnance, after that he hath received his charge at the councel's hands, he must first of all, in anywise before he shall go fourth to the camp, see that they lack no kind of ammunition, or such other necessaries which appertain to the said master of the ordnance.

"And there are appertaining to the master of the ordnance a lieutenant, and certain clerks which are all on wages.

"Also the said master of the ordnance must also receive the ordnance, shot, cornpowder, serpentine powder, match, and all other ammunition, as fire works, bows, arrows, strings, pikes, bills, halberts, harquebusses, lances, light horsemens staves, javelins, and bore spears.

"And further, the said master of the ordnance must receive all kinds of necessaries, that is to say, ladders, ladles and sponges for artillery; mattocks, spades, shovels, pickaxes, crows of iron, cart wheels for ordnance, carriages for ordnance, axletrees, hand axes, axletrees for ordnance, windons[1] for the defence of ordnance, cart traces, with all

1 *Windon.* From the context it is evident that the *windon* was what is now called a "chevaux-de-frise." The word has a great similarity to the word "windle," which in Lancashire signifies a machine or wheel on which yarn is wound, and which much resembles the modern chevaux-de-frise.

kind of cart wares, as ropes, cressets[1] and cressettes, lights, lanthorns, candle and links, with all other necessaries which must be foreseen, that there be no lack before their going on.

"Further, it is the office of the master of the ordnance, after he comes into camp, and the provost-marshall hath appointed the ground most mete and necessary for the artillery, then must the aforesaid master of the ordnance cause the said ordnance to be brought to the said place appointed, there to be placed to the most advantage.

"*Item.*—The said master of the ordnance must cause the said ammunition to be brought to the place appointed, and mete therefore, which must be trenched about, for the danger of fire; and the aforesaid master of the ordnance must charge some discreet man with the watch if it stands in need.

"Also the said master of the ordnance must see that there be attending on the office of ordnance certain artificers, as carpenters, wheelwrights, smiths, boweyers, fletchers, masons, and such other necessary men, mete and convenient therefore.

"The said master of the ordnance his office is, that if there be any captain that lacketh ammunition for his soldiers, the said captain shall come to the master of the ordnance, and he must command the clerk of the ordnance to deliver such ammunition as he lacketh; providing always that the clerk of the ordnance do take a bill of the captain's hand, or of his lieutenant for the said ammunition, and at the pay day the clerk shall deliver the said bills unto the treasurer, that he may stay so much money in his hands as shall answer the Queen for the ammunition so delivered.

"Furthermore, it is the office of the master of the ordnance that if the enemy and you join battle the ground being appointed by the officer of the field, where the battle shall be pitched, to repair to the field, there to see the ordnance planted to the most advantage; and if occasion shall be given, to remove the said artillery as shall seem good to the master of the said ordnance, and in anywise to circumspect that the master-gunners do their duties belonging to them."

---

1 "*Cresset.* An open lamp, suspended on pivots on a kind of fork, and carried upon a pole, formerly much used in nocturnal processions. The light was a wreathed rope smearod with pitch or rosin, stuck on a pin in the centre of the bowl. The cresset was sometimes a hollow pan filled with combustibles, and indeed, any hollow vessel employed for holding a light was so called."—*Dictionary of Archaic and Provincial words.*

"The *cresset-light* was set upon beacons and watch towers, or carried upon a pole. It was a square or circular framework of iron, having open ribs or hoops in which pitched ropes or other combustible materials were burned."—Knight's *Pictorial Shakespeare.*

* * * "At my nativity,
The front of heaven was full of fiery shapes,
Of burning cressets."—1 Henry IV. Act. 3, Scene 1.

Cressette is the diminutive of *cresset.*

Another and more ancient manuscript from the same author, Vol. I. p. 202, gives several curious particulars respecting the power and perquisites of the master of the ordnance, intermixed with the duty of the provost-marshal of the artillery under the following head:—

"These be the authorities and power that the PROVOST-MARSHALL and his lieutenant have in the jurisdiction of the artillery,—

"First, the provost-marshall hath none authority to bear his staff, nor his lieutenant within the jurisdiction of the artillery, without lisence of the provost of the artillery, but let his staff before the artillery gate, as the antient custom is in the realms of France, Spain, Portugal, Naples, and Le Ciceley and Levant.

"*Item.*—If there be any person found in the artillery charged with a crime, so must the provost of the artillery deliver him out of the artillery unto the provost marshall or his lieutenant, reserving always that the said provost of the artillery shall keep for himself all those goods and clothing belonging to the aforesaid 'crymeneux dedely patient.'

"*Item.*—All those of the small artillery, as serpentines, courtoux, bombardes, are bounden, and must foarth with each of their master-gunners, at the commandment of the original master-gunner, upon the pain and correcting of the chief master of the artillery and his councel.

"*Item.*—That all the carpenters are bounden to be by their mantells and works in the artillery, as well in the field as elsewhere, that is in any business to do, upon pain to abide the correcting of the said master and his counsel.

"*Item.*—The master of the artillery shall do cry with sound of trumpet within his jurisdiction of the artillery, with his provost, that all master gunners of courtoux, or serpentines, and all other being of the same offices, that each man shall keep the ordinances made by the great master of the artillery, every man severally keeping his place, his piece, and their fire and powder; and their servants and boys shall diligently watch upon their masters, and abide by them, to see what they have need of, or anything should lack, as his powder, stores, pellets, necessary unto them where they lie, upon the pain to abide the correction of the master of the artillery, lieutenant, or provost.

"*Item.*—The provost shall go with the lieutenant of the king or prince of the army, with the consent and license of the great master of the artillery, to make place, as is accustomed to be done of old, and that they shall take footman enough to make a place to shoot, and ditch it as appurtainith, within the which they may bring in their waynes, and carts, with powder and other necessary things, and so thereupon to depute and order in six or seven men deputed or assigned by the master of the artillery, to the defence of the same, upon the pain to be corrected as is aforesaid.

"And whereas the master of the ordnance is committed and made by any king, prince, or captain-general, and by their counsel, is admitted and charged with the gunners in town or in field, there ought no man without commandment of the said prince, lieutenant, captain-general, and the said counsel, to put no gunner, in or out the said ordnance, without the license of the said master, or the lieutenant for his discharge.

"*Item.*—All other waynes and carts that be laden shall be set in good ordinance as it hath been of old and antient custom to be, on pain as is aforesaid.

"*Item.*—That all the mantells and timber work, basilisques, water mills and other instruments belonging to the sieged town or castle, the which shall be brought so secretly by night and darkness as is possible to be done.

"*Item.*—The gentlemen deputed to give attendance upon the master of the artillery, to govern any bombards or cannon, shall not do nothing otherwise then is ordeyned by the said master of the artillery upon pain to abide the correction of the said master.

"*Item.*—That all servants and officers that have to do under the authority of the master of the artillery, and in his absence his lieutenant and officers, as his chaplain, receivor, comptroller, provosts, and clerks, master-gunners of cortolls[1] or serpentines, and all other servants, as waggoners, carters, their servants, with others shall keep and fulfill all such statutes as are ordeyned by the great master of artillery and his counsel, lieutenant and provost, upon pain to be corrected to the example of all others.

"*Item.*—As a town is won, whether it be by assault, per force, subtile practice, or by any other manner given up, be it town, castle, pile, church, or bastile, or fortress, the chief master of the artillery or his lieutenant, shall ordain, that the master-gunners and their companies shall have the best bell within that place so won, or the church wardens shall appoint or compound with the great master of the artillery and his counsel; and that to be reported by the provost of the artillery, and given knowledge to the lords and rulers of that place so won with the commons of the same, what that the master of the artillery, his counsel, and the master-gunners and their companies have determined and ordeyned, by a couverable and reasonable estimation, to see and know if the lords and commons will hold the ordenance and appointment made.[2]

---

1 Courtoux or courtaux.

2 It is believed that this privilege still belongs to the artillery of all armies. When the town of Fritzlar, in Germany, was captured, the inhabitants were compelled to ransom their bells by the payment of 5000 crowns to the officer commanding the British artillery. This was the only ransom demanded of the inhabitants.

"*Item.*—That all the butchers of the artillery shall flay their beasts without the precinct of the artillery, and that they grave and bury the filth of those beasts in the earth without the artillery, upon the pain to be, &c. &c.

"*Item.*—That all the horses and other beasts that be killed, or die on their own deaths, being carrion, the provost of the artillery must convey them out of the park of the artillery, for because of infection, upon the pain of being corrected by the master and counsel, or his lieutenant.

"*Item.*—The provost of the artillery shall have his right of the victuallers within his jurisdiction in likewise as the provost-marshall hath in the great army by estimation."—*Harl. MS.* 4685.

1578. Queen Elizabeth's annual expense for the ordnance as found in *E. Codice, MS.* Taken from a collection of curious papers published by Francis Peck, M.A., dated 1732.

	£	*s.*	*d.*		£	*s.*	*d.*
Master of the ordnance, fee	151	11	8	Master-gunner of England, fee	66	13	4
2 clerks { the one per diem	0	1	0	2 gunner soldiers, fee a-piece, per diem	0	1	0
2 clerks { the other, do.	0	0	8				
Lieutenant, fee	36	10	0	And each a gunner's room[1] per diem	0	0	8
Clerk	12	13	4	Gunner Smyth, fee „	0	0	8
Surveyor, fee	36	10	0	And a gunner's room at „	0	0	6
Clerk	12	13	4	Gun-stone maker, fee „	0	0	8
Keeper of the great storehouse, fee	50	0	0	And a gunner's room at „	0	0	6
Clerk	12	13	4	Saltpetre-maker, fee „	0	0	6
Keeper of the small storehouse, fee	40	0	0	Carpenter, do „	0	0	8
Clerk of the great storehouse, fee	50	0	0	Engineer or artificer, do „	0	0	8
Clerk of the small storehouse, fee	36	0	0				

109 inferior gunners or cannoneers, viz.—

Fifteen per diem	0	1	0	Eight, per diem	0	0	6
Twelve „	0	0	8	The rest „	0	0	4

Belonging to Fortresses. Berwick.

at	£	*s.*	*d.*	at	£	*s.*	*d.*
Master of the ordnance, fee	146	0	0	28 gunners of the old ordnance, fee	284	0	0
Muster-master or controller of the check, fee	64	0	0	42 gunners of the great ordnance „	580	0	0
Master-gunner, fee	40	0	0	Artificers in the office of the ordnance, fee	2998	19	0

In other castles named (but not copied).

5 master-gunners, each per diem	0	0	8	154 gunners, each per diem	0	0	6

1578. An estimate of the remains in the office of ordnance in 1578 (20 Elizabeth), with the value in ready money of the several titles ensuing, as well within the Tower of London as aboard the ships.

---

[1] This refers to the practice of maintaining a certain number of non-effectives to increase the salaries of the effectives.

1. Remaining in store within the Tower of London.

Cannons	18
Cannon pieces	1
Demi-cannon	11
Culverings	8
Demi-culverings	20
Sacres	11
Minions	8
Fawcones	7
Fawconets	20
In all	104

2. Remaining on board the ships.

Cannon pieces	24
Demi-cannon	36
Demi-cannon pieces	5
Culverings	76
Demi-culverings	118
Sacres	123
Minions	30
Fawcons	39
Fawconets	3
Fowlers, with two chambers a piece, and port pieces	17
In all	504

All which do weigh, by estimation xiii. iiii^{xx}. xv^{mll}. vi^{c}. ii. quarters weight, which being rated at £4 the C weight, with a M. xi^{c}. and lx^{li}. for their carriages; the furniture amounted in money to vii^{m}. vii^{c}. iiii^{xx}. vi.l.[1]

All which pieces aforesaid do remain as before, over and besides all those that have been uttered out of the stores from time to time, for the supply of the forts and castles along the sea coast, and other her majesty's places of strength and service, by sundry warrants from her majesty and the counsel.—

3. In the Tower,—

Cross-backed and iron shot, round of several heights	47000
And stone shot for cannon pieces, port pieces and fowlers	4500
In all	51500

4. Aboard the ships,—

Iron shot	100000
Stone shot	1300
In all	101300

Amounting in money to the sum of £5,475.

5. Powder and stuff for powder in the Tower,—

Corn and serpentine powder	55 lasts.
Saltpetre	10000 weight.[1]
Sulphur	20000 do.
Corn and serpentine powder (dimid)	lasts.

Amounting in money to the sum of £6,617 10s.

6. Small guns and artillery, munitions, and rich weapons.

(1) In the Tower,

Calivers	7,000
Daggs	5,000
Match	60,000 weight.
Bowes	8,000
Arrows	1[illegible],000 sheeves.
Morispikes	10,000
Billes	3,500

(2) Aboard the ships,

Calivers	320
Match	300 weight.
Bowes	380
Arrows	380 sheeves.
Morispikes	460
Billes	460

[1] *Sic in orig.*

(3) In the Tower, rich weapons, viz. armed pikes, halberts, partizaunts, javelins, boare spears, pol axes, amounting in money to £2,300. In all £18,877 13*s.* 4*d.* Summa totalis of the value of £38,876 13*s.* 4*d.* The remains £8,876 13*s.* 4*d.*

1583
28 Feb. A warrant from her majesty to the Earl of Warwick, master of the ordnance, to forward by sea a quantity of shot, powder, and other furniture, to Sir Simon Musgrave, master of the ordnance at Berwick, and the Borders of Scotland.—*Addl. MS.* 5754.

1585.
4 June. A warrant from her majesty to the Earl of Warwick, or the lieutenant of the ordnance, Sir William Pelham, and other officers of the ordnance, to forward supplies of ammunition and stores to Jacques Wingfield, Esq., master of the ordnance in Ireland.—*Addl. MS.* 5754.

Robert, Earl of Leicester, with forces and fifty gunners, embark for Flanders.

December. A warrant by the Earl of Leicester to —— Wilford, to serve in Ostend as captain of canoneers; he afterwards became celebrated for his defence of Leith in Scotland.—*Harl. MS.* 5753.

Letter from Sir Francis Walsingham to the Earl of Leicester.—"That the report of 100 pieces of brass ordnance, being missing through Sir William Pelham's neglect in overlooking his inferior officers, had like to have hindered his coming over into Holland to assist him with his advise, the Queen was so much offended with him."—*Harl. MS.* 285, 61.

1588.
July. The proportion of ordnance for the army appointed to be drawn together to encounter the enemy,—

	Ordnance.	Horses.		Ordnance.	Horses.
Demi-cannon	4	18	Sacres	3	7
Culverins	4	14	Falcons	2	3
Demi-culverins	3	12	Minions	2	6

To each of the same pieces so many horses.

The proportion or ordnance that is appointed for the guard of her majesty's person.

	Ordnance.	Horses.		Ordnance.	Horses.
Cannon	6	24	Demi-culverins	6	12
Demi-cannon	6	14	Sacres	6	7
Culverins	6	14	Minions	6	6

To each of the same pieces so many horses.

Harl. MS. 168, pp. 173, 174.

28 Oct. A warrant from her majesty to the Earl of Warwick, master of the ordnance, to forward supplies of ammunition and stores to Sir George Carew, master of the ordnance in Ireland.—*Addl. MS.* 5754.

Rates of entertainments of the officers of the companies for the service of the year 1588.

per diem.	£	s.	d.
Master of the ordnance..............	0	10	0
Lieutenant	0	6	8
Inferior officers of the ordnance :—			
10 laborers			
Muster-master	0	6	8
4 clerks at 2s. each	0	8	0
Commissary of victuals	0	6	8
1 clerk	0	2	0
Proof-master	0	6	8
Master of the carriages	0	4	0
Master-carter			

per diem.	£	s.	d.
Four clerks..............................			
Quarter-master	0	10	0
6 farriers................................			
Provost-master	0	6	8
2 tythers,[1] at 1s. 4d. each............	0	2	8
Judge-general	0	2	8
Entertainment of the officers of the regiment:—			
Colonel, being a nobleman	1	0	0
He being a private, or nobleman's son	0	13	4
Lieutenant-Colonel	0	6	8

Harl. MS. 168, p. 175.

1589. Ambrose Dudley, Earl of Warwick, master of the ordnance, died the 21st February.

"The Earl of Warwick, after his preferment to the office of the master of the ordnance, was besieged in the town of Newhaven in Normandy by Charles IX. of France. As he stood on the rampier at a parley, he was shot, contrary to the laws of arms, with a bullet impoisoned, which consumed his leg; and thereof after he had lived several years with great pain and impotensy, endured his leg to be sawed from his body, and died a few days after."—Thomas Louch's *Life of Sir P. Sidney*, p. 206.

A letter from Sir P. Sidney to the Lord Treasurer stating, "The queen at my Lord Warwick's request, hath been moved to join me in his office of ordnance, and as I learn her majesty yields gratious hearing to it." Dated 27th January, 1582.—*Harl. MS.* 6993, p. 471.

The appointment of master of the ordnance held by commission until 1596.

1590. 5 June. Her majesty's warrant to Sir Robert Constable, knight, lieutenant-general of the ordnance, to forward ammunition and stores to Sir George Carew, master of the ordnance in Ireland.—*Addl. MS.* 5754.

1594. Sir George Carew appointed lieutenant of the ordnance at £72 per annum; William Pastheriske, surveyor at £56 per annum. The appointments as follow were held by;—clerk of the ordnance Stephen Ridlesden at £71; storekeeper Thomas Bedewell at £40; clerk of the deliveries George Hogge at £18 5s.; clerk to the master-general, William Cudner £18 5s.; master-gunner of England, Stephen Bull at £70; keeper of the small guns Richard Paulfreyman at £50 per annum.

The master-gunner of England taught his scholars to shoot in great

1 "*Tith.* Tight or Strong."—*Dictionary of Archaic and Provincial words.*
Tythers. Strong men attached to the provost-master, to assist him in his duties.

ordnance at the artillery garden.—Taken from the *Ordnance Quarter Book* for October.

1595. William Lamon, proof-master, at £3. To Henry Pitt, Richard Phillips, Samuel Gaine, and Peter Gill for casting brass ordnance out of her majesty's metal. To Robert Goare, gun-stone maker, for several sorts of stone shot brought into her majesty's stores.—From the *Quarter Book.*

1596. Robert, Earl of Essex, appointed master of the ordnance. Sir George Carew, lieutenant of the ordnance with artillery embarked with the expedition to Cadiz, under the Earl of Essex, where they captured 1200 pieces of ordnance taken or sunk.

12 Feb. Stores and ammunition delivered from the Tower of London for Sir George Bourcher, master of the ordnance in Ireland. —*Addl. MS.* 5754.

1597. Sir George Carew, lieutenant of the ordnance, and Sir Christopher Blount as first colonel with artillery embarked with the forces under the Earl of Essex on an expedition to the Islands of the Azores.

John Lee appointed storekeeper to the ordnance.

John Linewray appointed jointly clerk of the deliveries with George Hogge.

Establishment of Artillery.

In the reign of Queen Elizabeth, especially in the year 1597, is as follows, viz.—

	£	*s.*	*d.*		£	*s.*	*d.*
Sir George Carew, knight lieutenant of Her Majesty's ordnance, for his allowance one quarter ...	18	0	0	Stephen Riddleshen, clerk of Her Majesty's ordnance, for his like quarter's allowance	17	12	0
				John Lee, keeper of H.M. stores, do.	10	0	0
William Parkeringe, surveyor of Her Majesty's ordnance, for his like quarter's allowance	18	0	0	George Hogge and John Lynewrace, clerks of deliveries	13	11	4
				William Oudner, clerk to Sir George Carew, Knight	4	11	2

Clerks daily attending in the said office for three months, at £5 each.

Richard Palfreyman.	Thomas Lemmas.
Edward Pasheringe.	John Squire.
William Scott.	Richard Lentall.
Richard Huynes.	That is, for one quarter's salary...£116 14 6

1598. John Davis appointed surveyor-general. He was knighted the following year.

1599. Names of the ordnance used in the navy. Taken from a return of the ordnance, dated 6th April, 1599.

	Bore, inches.		Bore, inches.
Cannon	8	Falconets	2
Demi-cannon	6¾	Port piece hall	
Culverins	5½	Port piece chambers	
Demi-culverins	4	Fowlers hull	
Sakers... Mynions	3¼	Fowlers chambers	
Falcons	2½	Curtalls	

With the army in Ireland under Lord Mountjoy there were 1600 gunners, cannoniers, armourers, and clerks of the ordnance.—GROSE'S *Military Antiquities*.

The Earl of Essex suspended from his office of master of the ordnance.

1601. John Linewray appointed surveyor-general. Robert Johnson appointed clerk of the deliveries.

James I. 1603. Charles Blount, Earl of Devon, appointed master of the ordnance.

John Pavye appointed clerk to the master of the ordnance.

1604. Sir Amias Preston, knight, appointed storekeeper to the ordnance.

The principal officers of the ordnance knighted in this year.

1605. Sir George Carew, lieutenant-general of the ordnance, created Baron Carew.

1606. Sir Roger Dallison appointed master-surveyor of the ordnance.

1607. William Bull appointed master-gunner of England. Francis Morris appointed clerk of the ordnance.

1608. Lord Carew appointed master-general of the ordnance; Sir Roger Dallison, knight, appointed lieutenant-general to ditto; Sir John Key, knight, appointed surveyor-general to ditto; Richard Paulfreyman, appointed clerk to the master-general.—*October Quarter Book*.

1609. Sir Roger Arconghilt appointed storekeeper to the ordnance. George Hooker appointed clerk to the master-general.

1611. William Hamond appointed master-gunner of England.

1612. Samuel Hales appointed storekeeper to the ordnance.

James Paulfreyman appointed keeper to the small guns and shot.

1616. Sir Richard Morrison appointed lieutenant of the ordnance. Nedtracy Smart appointed storekeeper to ditto.

The names of all the officers called MASTERS OF THE ORDNANCE in England, which was an office first created by Henry VIII.

Sir Thomas Seymour.	Robert Devereux, Earl of Essex.
Sir Phillip Hobbye.	Charles Blount, Earl of Devon.
Sir Richard Southwell.	Sir George Carew, Baron Carew of Clopton.
Ambrose Dudley, Earl of Warwick.	

The names of all the officers called LIEUTENANTS OF THE ORDNANCE in England, which was an office first created by Henry VIII.

Sir Christopher Morrys.	Sir William Pelham.
„ Francis Flemyng.	„ Robert Constable.
„ William Bromfield.	„ George Carew, Baron Carew of Clopton.
„ Edward Randolphe—in Ireland.	

Harl. MS. 5180, p. 83.

1618. Establishment of the train of artillery.

1 General of artillery.	1 Captain of miners.
1 Lieutenant of do.	25 Miners.
10 Gentlemen of do.	1 Captain of pioneers.
25 Conductors of do.	25 Pioneers.
1 Master-gunner.	1 Surgeon.
136 Gunners.	1 Surgeon's mate.
1 Petardier.	

SMITH'S *Military Dictionary.*

1620. Thomas Powell appointed storekeeper of the ordnance; Edward Johnson, ditto, clerk of deliveries; John Brown, master-founder of the iron ordnance, and Richard Phillips one of his majesty's founders for brass ordnance.

4 July. At Whitehall the 4th July, 1620.

Present.

Lord Privy Seal	Mr Secretary Maunton.
Earl of Arundell............	Mr Secretary Calvert.
Earl of Southampton	Mr Chancellor of the Exchequer.
Lord Digby	Master of the rolls
Mr Treasurer	Sir Edward Cooke.

"Whereas the Tower of London being his majesty's royal castle, one of the principal and most eminent forts of this kingdom, a great strength and ornament to the city, the chief storehouse and magazine of warlike provisions of this kingdom, hath antiently been fortified, not only within the walls, ditches and wharfs of the same, but also such care taken in the minorits,[1] and other neighbouring places, as well for the lodging and receipt of the principal officers of the ordnance, as likewise for artificers, gunmakers, wheelers and others, whose trades are appropriated to military ends, as nothing was there almost wanting, which was fit for the state to provide either for honour or safety.

"But for as much in these latter times, either through the evil example or tolleration of some lieutenants of the tower, or by abusing of

1 The Minories.

trust reposed in some officers who have particular relation to the place; there hath insencibly crept in diverse abuse and incroachments whereby the said antient limits of the tower, and those other habitations and storehouses appointed for public use are now perverted to private profit; the splender and magnificence of the said Royal Castle being by that means defaced, and the place itself as it were besieged in the wharf, ditches, and liberties thereof."

Upon complaint made of all which abuse the Board ordered as follows:—

"That whereas there was a lease procured from his majesty to William Hammond, master-gunner of England in the Tower, of the artillery yard near unto the minorits, and an antient obsolete name of the tifal yard for 223 years, and the same lease being defective and apparently void, and so acknowledged by the same Hammond and his council, for manifest imperfections found out by the said Sir Edward Cooke, was by the said Hammond delivered into the hands of the said Sir Edward Cooke, and now remaining in the councels chest.

"It is ordered that the said artillery yard be for the future restored to the public use for which it had been formerly employed, viz. for exercise of arms and artillery, and that the same shall not henceforth be alienated or converted to any other use:

"And whereas likewise there are leases procured from his majesty as well by as for Sir Richard Dallyson, Knight and Bart., late lieutenant of the ordnance, for 60 years of divers houses in the minorits, belonging to the office of ordnance, and therefore reserved for gunmakers, wheelers, and other artificers serving for the defence and safety of the realm. As also by Sir Thomas Mounson, late master of the armory, and likewise divers leases procured by divers persons, some of part of the tower wharf, others part of the tower ditch, the portcullis, and posterne gates, divers whereof as the said Sir Edward Cooke affirmed were void in law.

"It is ordered that as well the said houses belonging to the said office of ordnance and armory, as also the tower wharf, tower ditch, portcullis and posterne gates, be restored to their former use respectively and not alienated or converted to any other private use, upon any occasion for the future. And for as much as Sir Roger Dallyson and Sir Thomas Mounson, did when they were officers to the king, contrary to the trust in them reposed, procure their said estates to his majesty's prejudice and dis-service: it is ordered that they be both called to the Board, where the king's councel learned in the law are likewise to attend, and there they shall receive their lordships further pleasure, &c. &c."

1620. 4 July. Report made to the Prince's Higness in councel by Sir Richard Morryson, lieutenant of the ordnance, of the forts, castles, &c., on the sea coast, and a survey and remains of ordnance, shot, stores, &c. in the Tower of London with the names of the officers of the ordnance.

	Salary	£	s.	d.
The Lord Carew, master of his majesty's ordnance......				
Sir Richard Morryson, knight, lieutenant of	do.	76	13	4
Sir John Kay, knight, surveyor of	do.	36	10	0
Francis Morris, Esq., clerk of	do.	36	10	0
Thomas Powell, Esq., clerk of the stores		54	13	0
Edward Johnson, Esq., clerk of the deliveries...........		18	5	0

Other inferior officers and artificers.

	£	s.	d.		£	s.	d.
Gun foundery, 6	114	10	10	William Smeadon, wheelwright...	15	5	0
A founder of shot	9	2	6	A purveyor of arms..................	12	3	4
A stone shot-maker..................	12	3	4	Anthony Feraby, purveyor for timber	12	3	4
John Luggar, case maker	9	2	6	Lewis Tayte, the master-smith ...	9	2	6
Henry Rowland, gun maker	24	6	8	John Fletcher, rope and match maker	12	3	4
Another gun maker	12	3	4	Hugh Peice, porter of the minorits	12	3	4
A cross-bow maker	6	1	8	John Reynolds, master-gunner of England.			
John Jefferson, bowyer	9	2	6	7 gunners, double fees.			
John Powell Fletcher	9	2	6	13 do. .... each at	0	1	4
A bow-string maker	9	2	6	69 do. .... „	0	0	8
Matthew Banks, carpenter.........	12	3	4				

Brass ordnance mounted, serviceable.

Cannon of 7 and 8 inches	20	Port piece halll[1]	10
Cannon perriers	7	Chambers for them	44
Demi-cannon	16	Fowlers halls[1]	56
Culverins	36	Chambers for them	76
Demi-culverins....................................	18	Basse	2
Sakers	16	Rabbonets	18
Minions	13	Mortar pieces	7
Faucons	9	Bombard	1
Fauconetts	15	Pieces for shooting divers bullets	2

Cast-iron ordnance, dismounted.

Culverins	10	Minion	1
Demi-culverins	44	Faucon	1
Sakers	36		

With small guns, rich weapons, weapons for service, weapons for devastation, corn powder, match, round shot iron, cross-barred shot, chain shot, and Langrell cases with broad square, and Hayle shot, &c.[2]

"The state of the receipts and deliveries from the death of King Henry VIII., the 27th January, 1546, to the last day of December, 1561, Edward VI. and Queen Mary, wherein is remembered the whole mass and stores at this day remaining in your several armouries, and in the hands of your nobility and subjects :

"From Winchester-place in Southwark, then the house of the Marquis of Northampton,

Demi-lances.........	160	Corslets	2

From Puckeridge, in the county of Hertford, the armoury of Doctor Cranmore, late Abp. Almaine R——, steel saddles 180.

Hanse Hunter, the Queen's armourer, maces 457—Elizabeth.

---

1 *Sic in orig.*

2 Much of what follows is so badly copied in the MS. as to be nearly unintelligible.

Sir Richard Southwell, master of the armoury to Queen Mary, shaffrons 329.

Sir George Howard, master of the ordnance to Queen Elizabeth.

Mr William Winter, master of the ordnance to Queen Elizabeth.

The Lord Darcie, master of the ordnance in Edward VI.

James Fuller, yeoman of the armoury, Greenwich—Elizabeth.

Sir Thomas Gresham, factor for King Edward VI. beyond the seas.

Corslets leather, barbs aynetts. One Sir Henry Palmer, kt., master of the ordnance at Bolloign, 4th Edward VI.

Remains in the Tower armoury. Westminster armoury.

Hampton Court armoury."

" Marrions collars vambrace, 22 pair gantlets, 32 pair leather barbes, sallets, and skulls, shirts of mail, sleeves of mail, burgonets, collars anymes, targets, swords, paring staves, punching staves, counter burrs, harness for King Henry VIII. A brigandine for his own person, cannon perrier, bombard port pieces, morrice pikes, javelins, &c."

Clerks of the ordnance, anno 28, Henry VIII., as appear by old records in the office :—

Mr George Hagge.
Mr John Linewraye.
Mr William Hurley—the money.
Mr Anthony Anthony—the store.
Mr Rogers succeeded Mr Hurley.

In Queen Mary's time,—

Mr Painter possessed of a gunner's fee, became clerk of the ordnance by the friendship of Sir Richard Southwell, knight, master of the ordnance.

In Edward VI., when Sir Francis Fleming was lieutenant of the ordnance, the said Anthony was surveyor of the ordnance, and Brian Hogge clerk of the ordnance.

List of master-gunners of England:—

Christopher Gould.	
Richard Webb	Elizabeth.
Anthony Fourntler, a soldier	"
Stephen Bull	"
William Bull	
William Hammond	Elizabeth.
John Reynold	"
John Wornn, Esq., an experienced soldier, a Scotchman by birth. Fee per annum £36. 10*s*.	

Master-gunners out of the ordinary of the office,—

By title of proof-master for proving of ordnance powder, saltpetre, and match. 200 scholars sworn to the practice of shooting in great ordnance. For powder, shot, match, planks, and wadding. Spent in training of scholars at the artillery garden.

The yearly allowance due by presidents unto the office of ordnance, 1599.

	£	*s.*	*d.*
To the master of the ordnance			
To the lieutenant	72	0	0
To the surveyor	56	0	0
To the clerk of the office	69	0	0
To the clerk of the deliveries	18	5	0
To the keeper of the stores	26	8	0

The copy of Mr Linewraye's book of reformation of abuses in the office of ordnance.

1st. False receipts charged upon the keepers of the stores,—

One Painter was convicted in the exchequer for the sum of £500.

The great gally, Sir John Luttrel, captain of the same.

This indenture made at the Tower of London the 31st June, in the 36th year of the reign of Henry VIII. between Sir Thomas Seymour, knight, master of the king's ordnance, and Thomas Serjeant, master-gunner of the said ship.

Sir Christopher Morrice, kt. Henry VIII., per annum £66. 8s.; he lies interred in St Peter's Church in Cornhill, by the title of master-gunner of England to Henry VIII.[1]

Sir Francis Fleming, kt., 1553, tempore Henry VIII. Edw. VI.

March. Sir Robert Southwell entered the 26th February, 1st Queen Mary.

Sir —— Bromfield at Newhaven, 4th Elizabeth 1563.

Edward Randolph, Esq., in Ireland 1556, the 8th Elizabeth, lieutenant of the ordnance.

Sir William Pelham died in Ireland 1587, or in Flanders the 19th January; he lies interred in the minorits church.

Sir George Carew, kt., from 1592 to Midsummer 1608.

Sir George Dallysore, from 3rd January, 1607, finish 1608 and 1615.

Sir Richard Morryson, sen., from Midsummer 1615, the 4th Jacobi.

Sir William Harrington.

Sir William Heydon, kt., Midsummer 1626, drowned in the voyage with the Duke of Buckingham to the Isle of Rhee.

Captain John Heydon after him, 1627.

November 23, 1567. For removing of powder out of the vault in the White Tower. Mr. Richard Webb, master-gunner for 24 days £1. 4s.

In 1568 I find charges for casting leaden bullets for carriers and culverins; when drawing and removing of powder out of one place to another, when mounting of ordnance; so that it appeareth in those days, labourers were not trusted about the powder, &c.—*Harl. MS.* 5912.

1620. A King's Warrant and a list of the Royal Army intended to be raised for the recovery and protection of the Palatinate, consisting of 25,000 foot, 5000 horse, 20 pieces of ordnance and artillery, with their officers, artificers, and attendants, ammunition, stores, pay, &c.

Nature of ordnance.

		No.
With block carriages.	Mortars, brass ...	2
	Cannon royal ...	4
	Demi-cannon ...	4
	Culverins	6
	Demi-culverins...	4
	Sakers	2

	Per diem. £	s.	d.
The master of the ordnance ...	3	0	0
The lieutenant....................	1	0	0
A Surveyor	0	6	8
2 clerks, each	0	2	0
1 auditor	0	6	8
2 clerks, each	0	2	0

1 *Vide* STOWE's *Surry*, folio 196 and page 211, Morrice. Yet in the chronicle I find that in 1544 his title was master of the ordnance.—*Cleaveland MS.*

	Per diem. £	s.	d.
Aray master	0	6	8
2 clerks, each	0	2	0
A convoy-master-general	0	10	0
6 engineers for fortifications, each	0	6	8
1 clerk and 3 conductors of the works, each	0	2	6
The clerk of the ordnance	0	6	8
4 clerks, each	0	2	6
9 gentlemen of the ordnance, each	0	3	0
30 harquebussers, each	0	1	6
A quarter-master	0	5	0
8 labourers, each	0	0	10
4 horsemen, to attend the quarter-master, each	0	1	6
3 farriers or labourers under him, each	0	2	0
A commissary of victuals	0	5	0
2 clerks under him, each	0	2	0
A commissary for the horses, carts, and powder	0	6	8
2 clerks to attend him, each	0	2	0
A purveyor-general, both for the numerous victuals, and all other necessaries belonging to the ordnance	0	6	0
Yeoman allowed him	0	1	6
A master of the carriages for the artillery	0	6	0
2 halberdiers to attend him, each	0	1	0
The master of the miners	0	5	0
25 other miners, each	0	2	0
3 conductors of the carriages, each	0	3	0
3 captains (to 450 pioneers), each	0	4	0
3 lieutenants, each	0	2	0
3 conductors of the pioneer works, each	0	1	6
2 chief petardiers, each	0	6	8
To every of the 10 attendants, each	0	1	6
1 master-gunner	0	6	0
3 master-gunner's mates, each	0	2	6
3 constables or quarter-gunners, each	0	2	0
124 other gunners, each	0	1	6
212 labourers, each	0	1	0
1 provost-master of the artillery	0	5	0
3 provosts, each	0	1	6
3 under do., each	0	1	0
1 founder of brass ordnance	0	3	0
Man at	0	2	0
Master of the fireworks	0	4	0
1 drummer	0	1	0
1 trumpeter	0	2	0
1 barber-surgeon	0	2	6
2 under barber-surgeons, each	0	1	0
1 master-carpenter	0	3	0
2 mates, each	0	2	0
24 other carpenters, each	0	1	6
1 master-smith	0	3	0
2 mates, each	0	2	0
18 servants, workers for 3 forges, each	0	1	6
1 master-cooper	0	3	0
2 mates, each	0	2	0
18 servants, each	0	1	6
1 farrier	0	2	6
6 servants, workmen, each	0	1	6
450 pioneers, each	0	0	8
65 conductors, each	0	2	0
3 tent-makers, each	0	2	0
9 servants to them, each	0	1	0
1 armourer	0	3	0
4 servants to him, each	0	1	6
1 collar-maker	0	3	0
4 servants to him, each	0	1	6
200 horses, each	0	1	0
100 carters, each	0	1	0

Harl. MS. 5109.

1623. John Reynolds appointed master-gunner of England.

30 July. A survey made of the castles, forts, and places, found placed on either side of the river Thames and Medway; and also of the castles, forts, with all their fortifications along the coasts of Kent, Sussex, Hampshire, and so onwards to the south coast, west, and to the Land's End, including the forts at Cornwall, by Sir Richard Marryson, kt. lieutenant of his majesty's ordnance, Sir John Ogle, kt. and colonel, Sir John Kay, surveyor of his majesty's said ordnance, in pursuance of certain instructions from the lords higness most honourable privy council.—*Harl. MS.* 1326.

Charles I. In Charles I. reign, Vicount Valentia, the master-general of the ordnance, was joined in a commission as councel to the Duke of Buckingham, when he commanded in chief the fleet.

1625. Sir William Harrington appointed lieutenant-general of the ordnance.

Sir Alexander Brett, appointed surveyor-general.

Lord Carew, master-general, created Earl of Totness.

1626. Sir William Heydon, kt., appointed lieutenant-general of the ordnance.

King Charles granted a patent for fourteen years to Arnold Rotispen, for making guns of all sorts, both great and small, after a new way or manner, not formerly practised by any within these dominions.—RYMER.

Sir William Heydon, lieutenant-general of the ordnance was drowned on the landing of the expedition under the Duke of Buckingham at the Isle of Rhee.

1627. 1 Oct. Sir John Heydon appointed lieutenant-general of the ordnance.

Sir Paul Harris, bart., appointed surveyor-general.

1628. Paid to Sir Sackville Trevor, kt., for the price or value of one culvering of brass, which was taken in the French prize, called "St Esprit," and allowed unto him by an order from the lords of his majesty's privy council, the sum of £208. 6*s*. 8*d*.

1629. The British artillery were employed and extremely active at the famous siege of Bois le Duc, which commenced in May under the Prince of Orange, and surrendered the 14th September.

1630. Lord Vere of Tilbury appointed master-general.

18 Nov. Commissioners and assistants to them for the office of ordnance.—

This day came into this office, the Lord Vic Wimbleton, Vic Dorchester, and Mr Secretary Cook, 3 of the commissioners for this office, and Sir Henry Spiller, Sir Kellan Digby, and Mr Bingley, assistants to the said commissioners and presented their commissions.—*Harl. MS.* 422.

1631. Francis Coningesby appointed surveyor-general.

1632. The British artillery employed at the siege of Marstrickt with the army under the Prince of Orange, which capitulated the 24th August.

1633. George Clerk appointed clerk of the deliveries.

1635. Mountjoy, Earl of Newport, appointed master-general; Richard March appointed store-keeper; Thomas Barnard appointed clerk to the master-general.

20 Mar. Commissioners for the ordnance and gunpowder,—

Earl Marshall, the Earl of Arundell.	Lord Cottington.
Earl of Dorset.	Sir John Coke, Kt.
Earl of Newport.	Sir Francis Windebank, Kt.
Lord Vic Wimbledon.	

A warrant to grant from time to time such powder as the master of the ordnance may require, dated the 20th March in the 10th year of our reign, 1635.

1636. Edward Sherburne appointed clerk of the ordnance.

1638. King Charles directed a cannon to be cast in this year, which was placed in St James's park, it was emphatically called "The Gun," bearing this inscription,—

> Carolus Edgœri sceptrum stabilivit aquarum.
> "The sceptre of Edgar established on the waters by Charles."

N.B.—The gun has since been removed to the Repository at Woolwich, where it now (1845) remains.

Extract from March *Quarter Book*.—The Earl of Newport, master-general; Sir John Heyden, lieutenant-general; John Reignolds, master-gunner of England.

June *Quarter Book*.—James Weymes, master-gunner of England.

1639. A list of the train of artillery, according to his majesty King Charles I. directions, reduced to such a number of officers and other ministers as will be merely necessary for a train of 30 or 40 pieces of ordnance for the army against the Scotch :—

	Per diem. £	s.	d.
The general of the ordnance	4	0	0
Lieutenant	1	0	0
A comptroller	0	10	0
2 commissaries of the two magazines of ammunition, viz.:— the train...	0	6	0
the army...	0	5	0
4 clerks under them, each	0	2	0
1 engineer	0	8	0
ditto	0	6	0
2 clerks for them, each	0	2	0
6 conductors of the trenches and fortifications, each	0	2	0
1 fire worker	0	3	0
His assistant	0	1	8
1 petardier	0	2	6
12 assistants, each	0	1	0
1 master-gunner	0	6	8
4 gentlemen, each	0	4	0
Master-gunner's mates, each	0	2	6
30 gunners, each	0	1	6
A paymaster	0	5	0
Captain of the pioneers	0	5	0
Quarter-master	0	4	0
4 conductors of the mattrosses, each	0	2	6
40 mattrosses, each	0	1	0
A purveyor	0	3	0
1 master-smith	0	3	0
6 servants under him, each	0	1	0
1 master wheelwright	0	2	6
4 servants under him, each	0	1	0
1 tent maker	0	2	8
2 servants, each	0	1	0
1 tent keeper	0	1	6

	Per diem £	s.	d.
1 assistant to him	0	0	8
1 master-carpenter	0	3	0
6 servants under him, each	0	1	0
1 cordage maker	0	2	0
2 servants under him, each	0	1	0
4 armourers, each	0	2	6
4 servants under them, each	0	1	0
2 gun-smiths, each	0	2	6
4 servants, each	0	1	0
1 harness maker	0	1	6
2 servants under him, each	0	1	0
1 farrier	0	2	6
2 servants under him, each	0	1	0
1 bridge-master	0	2	6
6 servants, each	0	1	0
1 provost-marshal	0	2	0
2 servants under him, each	0	1	0
1 surgeon	0	4	0
1 servant under him	0	1	0
1 wagoner for the train	0	5	0
1 assistant to him	0	2	6
1 saddle-maker	0	1	6
1 servant under him	0	1	0
1 cooper	0	2	0
2 servants under him, each	0	1	0
2 principal conductors, ammunition of the army, ditto of the artillery, each	0	3	0
40 conductors, twenty for the wagons, twenty for the ordnance, each	0	2	6
Commissary for the draught horses	0	4	0
2 assistants to him, each	0	2	6

GROSE'S *Military Antiquities*.

1640. Thomas Eastwood appointed clerk of the deliveries.

Sept. A list of such officers and ministers belonging to a train of artillery as are necessary, and now resident in London, viz.—

As comptrollers of artillery at 8*d*. per diem,—

Captain Younger, whose payment according to agreement with Sir Nicholas Biron, who took him from his employment in the Netherlands, was stated upon the lists of the Lord Marquis his troops at 15*s*. per diem, and remaineth amongst those that are still continued in pay, which he acknowledges to have received from Captain Vernon by the Lord Marquis his appointment until the 18th August last. But his lordship being in the north (whither he was directed and intended to have repaired) the lord marshal was pleased to stay his journey, and to command his attendance here upon the intended train of artillery. In which respect he humbly prayeth some speedy reward may be taken for the assignment of his payment here.

Gentlemen of the ordnance allowed 4*s*. per diem as appear by his majesty's list,—

Henry Bludder }
Edward Danker } For gentlemen of the ordnance.

Both formerly of the Lord Marquis' train, each 2*s*. per diem.

Battery-masters, one at 5*s*., the other 4*s*. per diem,—

Nicholas Samson, fit for a battery-master, formerly a serjeant under the Lord Marquis at 1*s*. 6*d*. per diem.

Conductors of the trenches, each 2*s*. per diem,—

Richard Huckley for a conductor of the trenches as serjeant at 1*s*. 6*d*.

Conductors to the train, each at 2*s*. 6*d*. per diem,—

William Griffith }
Clement Lynne } For conductors to the train of artillery as formerly to the Lord Marquis at 1*s*. 6*d*. each.

Harl. MS. 6851.

1642. His majesty directed Captain Green, comptroller of the artillery to join him.

22 May. His majesty's warrant appointing Lord Percy general of the train of artillery with the army.—*Harl. MS.* 6804.

20 Aug. Extract from the Journals of the House of Lords,—

Lord Wharton, speaker.

The House was informed, that Mr. March, one of the officers of the ordnance, was required, from the Earl of Essex, lord-general, to deliver the keys of the office of the ordnance, that ammunition might be had there, for the present service of the king, kingdom, and parliament; who thereupon petitioned the said lord-general as followeth, viz:—

To the right honourable Robert, Earl of Essex, his excellency.

The humble petition of Richard March, keeper of his majesty's stores.

Shewing,

"That your lordship's petitioner, having this present day received a warrant from your excellency, for delivering up the keys of the stores of arms remaining in his custody, doth, in all humbleness, profess his willingness to obey your lordship's commands, but, for your lordship's better satisfaction herein, humbly prays that your lordship would be pleased to take into your honorable consideration the petition annexed, prepared by the petitioner and other the officers of the ordnance, for their enlargement, wherein the petitioner and they become humble suitors for your lordship's honorable favour.

"And the petitioner shall pray,
for your lordship,
"RICHARD MARCH."

A petition of other officers of ordnance, to the lords of parliament to the same effect.

To the Right Honourable the Lords and Peers now assembled in Parliament.

The humble petition of Captain Francis Conningsby, Richard March, and Edward Sherburne, officers of his majesty's ordnance,

Shewing,

"That whereas your lordships petitioners have been divers days under restraint by your lordships commands, for not giving consent to the issuing of some munition in their custody; the petitioners humbly pray, that your lordships will be pleased to take into consideration.

(1) "That they are strictly bound, by the duty of their place, not to dispose of the munition in their custody contrary to his majesty's pleasure.

(2) "That before the petitioners received any commands from the right honourable the Earl of Essex his excellency, for the delivery of the munition in the warrant expressed, the petitioners received strict command, under his majesty's sign manual, not to issue out any munition without express warrant from his majesty.

(3) "That the petitioners have, these four years and a half, been unpaid the fees, and allowances belonging to their places.

"Therefore their humble suit is, that in regard to the giving of their consent to your lordships commands in this particular, would not only be a great breach of trust in them, but tend to their undoing; that your lordships would be honorably pleased to accept of their willingness and cheerfulness to obey your lordships commands, so far as may stand with their own integrity and safety, and that your lordships would be pleased to release them from their present restraint whereby they are put to great charge, and suffer much in their own private occasions.

"And your petitioners shall ever pray,
for your lordships,

"FRANCIS CONNINGSBY,
RICHARD MARCH,
EDWARD SHERBURNE."

It is ordered, by the lords and commons in parliament assembled, that the officers of the ordnance in the Tower of London shall forthwith, upon sight of this order, deliver the keys of the office of the ordnance, arms, ammunition, and stores there, to such as the committee for the defence of the kingdom shall appoint to receive them; or else that the doors of the said office shall be forthwith broken up, and the charge and keeping of the said arms shall be committed into the hands of such as the said committee shall think fit; who shall take inventories of the same, to the intent that a true account may be taken of the said arms to the use of his majesty, the parliament, and the kingdom.

1642. 22 Oct. His majesty's warrant for a proportion of ordnance to accompany 4000 foot.

Brass ordnance	Demi cannon	2
	Culverings	2

with their ammunition, materials, and attendants.

Captain Younger.........Comptroller	The master-carpenter
Monsieur Shebith.........Engineer	6 carpenters
Bodewinde De Hayne.	3 or 4 wheelwrights
Mr George Cole .... / Mr Matthew Wharton — Clerks of the munition of the ordnance and foot	3 coopers
	A ladle-maker
	A smith and his servants
Henry She .... / William Sneddall ...... — Gentlemen of the ordnance	Henry Elsey, and all the matrosses
	All the pioneers
Wm. Betts to be the chief gunner, and 8 gunners more, whereof Richard Bury to be one	6 conductors
	A fireworker

All the aforesaid officers, artificers, &c. are to be selected, and made choice of by Captain Younger, Comptroller.

Mem. This proportion assigned for Banbury but countermanded, in regard the whole train the next morning marched to Edghill.

22 Nov. At a counsell of war held at Reading, his majesty resolved that Sir John Heydon, kt., lieutenant-general of the ordnance, should send away to Oxford 8 pieces of ordnance, consisting of

Brass ordnance mounted,—

Culverings	2	6-lb. bullet...............	1
Demi-culverings	2	3-lb. bullet...............	1
12-lb. bullet	2		

with their ammunition material, and attendants,—

Gentlemen of the ordnance:—
Mr Stone
Mr Sherborne
8 Gunners
2 Conductors
1 Clerke
8 Mattrosses
8 Pioneers.

1643. Table, extracted from a work dedicated to the Duke of Buckingham by Robert Norton, engineer and gunner.

1 Names of the Pieces.	Height of Bore.	Length in Diameters.	Weight in Metal.	Weight of Powder.	Length of the Ladles.
	inches.		lbs.	lbs.	inches.
Cannon of 8	8	15	8000	40	24
Cannon of 7	7	16	7000	25	22
Demi-cannon	6½	18	6000	20	21
Culvering	5½	28	4500	15	20
Demi-culvering	4½	32	2500	9	18
Saker	3½	36	1500	5¼	16
Minion	3¼	30	1200	3¼	15
Falcon	2¾	42	700	2½	14
Falconet	2¼	48	500	1¼	12
Cannon-perior	9, 10, 12	8	3500	3, 3½, 4	3
Demi-can, drake	6½	16	3000	9	4½
Culvering drake	5½	16	2000	5	4½
Demi-culvering drake	4½	16	1500	3½	4½
Saker drake	3½	18	1200	2	4½

Earl of Peterburgh general of artillery; John White appointed clerk of the ordnance, John Falkener storekeeper to do; Stephen Darnelly Clerk of deliveries to do.—*Ordnance Quarter Book for January.*

Sir Timothy Tyrrell styled governor and general of the ordnance.

1 It may not be altogether out of place to make a few remarks on the "nomenclature" of the various species of artillery which are referred to in the foregoing paper. The word *artillery* is said by Moretti, in his "Treatise" (translated by Sir Jonas Moore, about 1683), to have been derived from the Italian word *artiglio*, signifying "the tallons or claws of ravenous fowls, perhaps because its shot flying far off dismembers and tears in pieces all that it meets." This derivation is probably correct, for we find that the various pieces of artillery have from very early periods been named after birds of prey and venomous serpents; thus, among the field pieces (which were necessarily more moveable than the others) we find the *smeriglio*, which is a long-winged hawk; the *falcon*; the *saker* (a species of falcon, and enumerated as such in the catalogue of AUCEPS in "Walton's *Angler*"). Again, among the heavier and longer pieces, we find the different varieties of the *culvering*, which name is derived from the *colubrine* or *colubrinetta*, a species of serpent: there is also the well-known *cannon serpentine*, as also the *cannon basilisk*. The *bastard* gun was one whose length was less than the *ordinary* of 32 calibres. The *moyenne* or *minion* (of 26 calibres) was a *bastard* piece. Moretti (before quoted) describes also *mortars* or *trabucchi*, "They are short pieces, of the nature of Petrieroes; and with these they shoot balls of stone, grenado shells, and cases full of small shot, not by a right line, but by a crooked from on high, so they fall where it should be appointed." . . . . . "There is no difference betwixt a *mortar* piece and a *trabucchio*, but in the placing of the trunnions." The *mortar* of the present day is the *trabucchio* referred to, while the ancient *mortar* is something like our "coehorn howitzer." The derivation of these names is as follows:—The *trabucchio* is derived from the Italian verb *traboccare*, to

John Pym, Esq. appointed lieutenant of the ordnance: he died the 8th December of the same year.

Sir David Walter appointed lieutenant-general of the ordnance.

George Taylor, surveyor-general to ditto.

1643. 30 Oct. Henry Lord Percy, Baron of Alnwick, master of the horse to his highness the Prince of Wales, and general of his majesty's train of artillery:

To Claud' Pelett de la Rinere, gentleman of the ordnance.

"By the power and authority given me from our sovereign lord King Charles under the great seal of England, as general of his majesty's train of artillery, I do hereby constitute and appoint you gentleman of the ordnance for his majesty's service; in which place you are to do and conform all things insident, and belonging to your said charge commanding those who it may concern to obey you as gentleman of the ordnance, according to this commission given you, and you are to observe and follow such orders, and directions as you shall receive from myself, and your superior officers belonging to the train of artillery according to discipline of war."

From Sir William Vavasor to the Lord Percy, general of the ordnance.

My Lord,

"We have taken Panswick with the loss of some men. and waist of much ammunition. The rebels have possessed themselves of many

---

throw, or hurl; while the word *mortar* has an evident affinity to the word *morte*, "death," or *mortorio, mortoro*, "a funeral."

The following is from the *Orlando Furioso* of Ariosto (circa A.D. 1560), Canto XI., Stanza xxi, et seq.:—

Orlando, I pursue,
That bore Cymasco's thunderbolt away,
And this had in the deepest bottom drowned,
That never more the mischief might be found.

But with small boot: for the impious enemy
Of human nature, taught the *bolt* to frame,
After the *shaft*, which after darting from the sky,
Pierces the cloud, and comes to ground in flame;
Who, when he tempted Eve to eat and die,
With the apple, hardly wrought more scathe and shame—
Some deal before, or on our grandsire's day,
Guided a necromancer where it lay.

More than a hundred fathom buried so,
Where hidden it had lain a mighty space—
The infernal tool by magic from below
Was fished and borne amid the German race,
Who, by one proof and the other, taught to know
Its powers, and he who plots for our disgrace,
The demon, working on their weaker coil,
At last upon its fatal purpose hit.

To Italy and France on every hand
The cruel art among all people past,
And there the bronze in hollow mould expand
First in the furnace melted by the blast;
Others the iron bore, or small or grand,
Fashion the various tube they pierce or cast,
And *bombard, gun*, according to its frame
Or *single cannon* this, or *double* name.

This *Saker, Culverine*, or *Falcon* hight,
I hear all names the inventor has bestowed;
Which splits or shivers steel and stone outright,
And where the bullet passes makes a road.
Down to the sword, restore thy weapons bright
Sad soldier to the forge, a useless load,
And gun or carbine on thy shoulder lay;
Who without these, I wot, shall touch no pay,

houses which will be taken by cannon, I must therefore desire your lordship to send me 20 barrels of powder more, with some hand grenades, and if you please to order a mortar piece, I shall be able to do his majesty better service, and your lordship will much oblige,

"Your humble servant,

"WILLIAM VAVASOUR."

I beseech your lordship to send some more cannon bullets.—*Harl. MS.* 4713.

1644. Sir Walter Erle appointed lieutenant-general of the ordnance.

Thomas Heselrige appointed clerk of the deliveries.

8 Apr. A warrant from his majesty to Sir John Heyden, kt. lieutenant-general of the ordnance, for paying the train of artillery.

25 Apr. His majesty's warrant to Henry Lord Percy, general of the artillery, or the lieutenant-general of the ordnance, or any other whom it may concern, to send from the magazines 150 shot for a minion piece, 2000 lbs. of match, and 2000 of bullets to the garrison in Reading.

8 Oct. To Sir Ralph Hopton, general of the artillery, and to the officers of the same, to deliver powder and bullets for the garrison at Milbrooke, and Mount Edgecombe.—*Harl. MS.* 6802.

1645. A list of the officers belonging to the train of artillery in the Parliamentary army, commanded by Sir Thomas Fairfax.

Lieut.-general Hamond ....lieut.-general of the ordnance.
Captain Dean..............................comptroller of the ordnance.
Master Hugh Peter ....chaplain to the train.
Peter Manteau Van Dalem ....engineer-general.
Captain Hooper ....engineer-extraordinary.
Eval Tencene ....chief engineer.
Master Lyon } Mr. Tomlinson } ....engineers.
Master Francis Favin ....master-gunner of the field.
Master Matthew Martin ....paymaster to the train.

Colonel Rainsborow
Lieut.-Colonel Bowen
Major Done }
" Crosse } slain at Sherburn
" C. Crosse }
" Edwards
Captain Edwards
" Drury
" Dancer
" Horsey, slain at Sherburn
" Creamer, wounded at Sherburn
" Sterne, slain at Bristol
Captain-lieutenant Flemming, slain at Sherburn
Colonel Weldon, now Lelburne
Lieut.-Colonel Kempson

Major Masters
Captain Peckham
" Fenton
" Franklin, slain at Exeter
" Holmes
" Dorman
" Tolhurst
" Munday, died in the west
" Welden
" Kaine
Master Phips, commissary of ammunition
Mr. Thomas Robinson, commissary of draught horses
Capt.-lieuts. {Desborow / Brent ...} of firelocks
Captain Cheese, of pioneers

1646. William Billers appointed clerk of the deliveries.—*Ordnance Quarter Book.*

1649. Richard Wollaston appointed master-gunner of England.—*Ordnance Quarter Book.*

1650. Major-general Harrison appointed lieutenant-general of the ordnance.

Letter from Cromwell to the speaker of the House of Commons, dated Edinburgh, 24th December, 1650.

"It having pleased God to cause the castle of Edinburgh to be surrendered into our hands this day about 11 o'clock, I thought fit to give you such account thereof as I could, and the shortness of time would permit. I sent a summons to the castle upon the 12th instant, which occasioned several exchange of returns and replies; which for their unusualness I also thought fit humbly to present to you. Indeed, the mercy is very great and seasonable. I think I need say little of the strength of the place, which, if it had not come as it did, would have cost very much blood to have attained, if at all to be attained; and did tye up your army to that inconvenience, that little or nothing could have been attempted whilst this was in design, or little fruit had of anything brought into your power by your army hitherto; without it, I must needs say, not any skill or wisdom of ours, but the good hand of God hath given you this place.

"I believe all Scotland hath not in it so much brass ordnance as this place. I send you here enclosed a list thereof, and of the arms and ammunition, so well as they could be taken on a sudden, not having more at present to trouble you with, I take leave, and rest,

"Sir,

"Your most obedient Servant,

"O. CROMWELL."

A list of ordnance, &c. in the Castle.

Brass Pieces.—5 French cannon, or cannon of seven, 9 Dutch half-cannon or 24-prs., 2 culverins, 2 demi-culverins, 2 minions, three 3-prs., 2 falcons, 28 drakes, called monkeys.

Iron Guns.—The great murderer called "Muckle Meg," 4 cannon, 10 drakes, called monkeys, 2 petards. About seven or eight thousand arms, between three and four score barrels of powder, and great store of cannon shot.

1651. At the surrender of Stirling Castle amongst the guns found were eleven leather guns, as reported by Colonel Moncke, lieutenant-general of the ordnance to the lord-general Cromwell.

1655. John Hooker appointed keeper of the small arms.

1656. Lewis Awdeley appointed clerk of the ordnance.

1660. Sir William Compton, kt. appointed master-general.

Colonel William Legg	ditto	lieutenant of the ordnance.
Francis Nicholls	ditto	surveyor-general.
Edward Sherburne	ditto	clerk of the ordnance.
Richard March	ditto	storekeeper to do.
George Clerk	ditto	clerk of the deliveries.
Matthew Baylie	ditto	clerk to the master-general.
Colonel James Weymes	ditto	master-gunner of England.

6 Nov. In the House of Commons, Sir William D'Oiley reported from the committee for disbanding the army, what progress had been made in that service, in which the two following items as belonging to the ordnance are mentioned, viz.—

(1) *In England*, general officers with the train £1642. 13*s*. 6*d*.

(2) *In Scotland*, general officers and train £797. 11*s*. 3½*d*.

1661. Sir Bernard de Gomme chief engineer.

52 gunners	3 gunners at 1*s*. 1 do. 8*d*. 48 do. 6*d*.	each per diem.	paid in the March Quarter Book at the Tower.
1662.	13 gunners at 1*s*. 5 do. 8*d*. 72 do. 6*d*.	each per diem.	paid in the December Quarter Book at the Tower.

28 Aug. Warrant from the Earl of Peterburgh appointing Captain Martin Beckman, engineer-general of Tanger.

1664. 21 Oct.

John Lord Berkeley. Sir John Duncombe, kt. Sir Charles Chicheley, kt.	the appointment of the master of the ordnance held by commission.

Captain Martin Beckman appointed fire-master; Richard Batchler appointed keeper of the small arms; 98 gunners paid at the Tower.—*In December Quarter Book.*

1666. Captain Valentine Pyne appointed master-gunner of England.

1668. Thomas Benet appointed clerk to the master-general.

Charles II. 1669. 18 Mar. Nature of ordnance in the service, taken from an ordnance return, shewing the remains at the Tower, and the several magazines, castles, and forts,

Brass Ordnance.

Cannon of 8	Sakers
Cannon of 7	Mynion
Demi-cannon	3-prs.
24-prs.	Falcon
Culverings	Falconett
12-prs.	Brass baces of 7 bores
Demi-culverings	Inch and $\frac{1}{4}$ bore, and 7 other sizes
8-prs.	
6-prs.	

Iron Ordnance.

Cannon of 7	6-prs.
Demi-cannon	Saker
24-prs.	Mynion
Culvering	3-prs.
12-prs.	Falcon
Demi-culvering	Falconett
8-prs.	Rabonett

Brass Mortar Pieces.

$18\frac{1}{2}$-inches	$7\frac{3}{4}$-inches
$16\frac{1}{2}$ "	$7\frac{1}{4}$ "
$13\frac{1}{4}$ "	$6\frac{1}{2}$ "
9 "	$6\frac{1}{4}$ "
$8\frac{3}{4}$ "	$4\frac{1}{2}$ "
8 "	$4\frac{1}{4}$ "

Iron Mortar Pieces.

$12\frac{1}{2}$ "	$4\frac{1}{2}$ "

Harl. MS. 4244.

1669. Dec. Jonas Moore appointed surveyor-general of the ordnance.

1670. 4 June. His majesty's warrant, appointing Sir Thomas Chicheley master of the ordnance, with the profits of the office, &c. in as full a manner as Ambrose late Earl of Warwick, or Sir Phillip Hoby, kt. or Sir Richard Southwell, kt., Sir Phillip Sidney, kt., Robert late Earl of Essex, Charles late Earl of Devon, George Lord Carew, Howard Lord Vere, Montjoy Earl of Newport, or the said Sir William Compton, or any other persons, &c.

Richard Beck appointed clerk to the master-general.

George Wharton appointed by king's warrant a clerk of the deliveries of all, and all manner of ordnance, artillery, ammunition, and other necessaries, &c., and enjoy the same in as ample manner as Bryan Hogg, George Hogg, John Lynwray, Sir Robert Johnson, Edward Johnson, Henry Johnson, Thomas Eastwood, and George Clarke, or any of them, &c., &c.—13 March. *Harl. MS.* 4713.

7 Dec. David Walter, Esq. appointed lieutenant of the ordnance.

Samuel Fortsey appointed clerk of the deliveries. 103 gunners paid at the Tower.—*December Quarter Book.*

George Wharton appointed treasurer and paymaster, it being improper for the lieutenant-general as a military officer to be treasurer, and receive poundage out of the large sums that would then pass his hands. Before this time the master-general and principal officers had the power of selling the places under them, and claimed the old guns and stores as their perquisites.

1671. His majesty granted the old store-house and wharf at Woolwich, called the gun wharf, to Sir William Prichard for the sum of £2957.[1]

Ernest de Reus, fire-master to the office of ordnance.

The gunners paid quarterly at the Tower as the previous year.

Charles Beaumont appointed clerk to the master-general.

1671. 4 Sept. An account of the natures, weights, and lengths of 41 pieces of new iron ordnance, being weighed, and proved, from Mr Westerne, one of his majesty's founderers.

Cannon of 7, of 10 foot.

cwt.	qrs.	lbs.
63	3	21
64	0	21
62	2	7
62	3	7
62	0	21
65	2	21
61	0	21
63	3	7
62	3	21
62	1	21
63	1	21
61	0	21
62	3	7
64	2	21
63	1	14
62	2	21
63	2	7
64	0	21

Ditto of 9¼ foot.

cwt.	qrs.	lbs.
60	1	0
59	1	21
59	3	21
59	0	21
57	3	21
59	0	7
59	3	21
54	0	21

Culverings of 10 foot.

cwt.	qrs.	lbs.
43	2	7
44	1	21

Demi-culverings of 10 ft.

cwt.	qrs.	lbs.
36	2	7
36	0	21
36	2	21
36	1	21
36	0	21
36	2	14
29	3	21
36	0	3
36	1	21
35	0	21
35	2	14
37	0	21
36	0	21

1672. Sir George March appointed storekeeper to the ordnance.

Estimate of the storehouse to be set up at Woolwich for receiving fire-works. The building to be 70 feet in length by 20 wide, with doors, locks, and all materials £125.

An estimate for the carriage of guns, carriages, shot, &c., from the office of his majesty's ordnance at the Tower place, near Woolwich.

Loading and drawing shot from the old yard to the new, per load 1*s*. 4*d*.

Drawing guns from the old yard to the Warren	Guns	30 cwt. and upwards......	4*s*. 0*d*.
		30 „ and under.........	2*s*. 6*d*.

---

[1] This appears to be the first mention of Woolwich in these notes. For the History of the Royal Arsenal at Woolwich, see a paper by Lieut. Grover, R.E., in Vol. VI. of the Proceedings of the Royal Artillery Institution.

1672. 2 Apr. The following warrant contains the first notice of bayonets being introduced into the English army.

"Our will and pleasure is, that a regiment of Dragoons which we have established and ordered to be raised in 12 troops of fourscore in each besides officers, who are to be under the command of our most dear and most intirely beloved cousin, Prince Rupart, shall be armed out of our stores remaining within our office of ordnance, as followeth; that is to say, three corporals, two serjeants, three gentlemen of arms, and twelve soldiers of each of the said 12 troops are to have, and carry each of them one holbard, and one case of pistols with holsters, and the rest of the soldiers of the several troops are to have and to carry each of them one matchlocke musket, with a collar of bandileers, and also to have and to carry a *bayonet* or great knife, that each lieutenant have and carry one partizan, and that two drums be delivered out for each troop of the said regiment, &c.

"By his majesty's command,

"ARLINGTON."

"To Sir Thomas Chicheley,
Master-general of the ordnance."

1673. His majesty's warrant appointing Sir Thomas Chicheley master of the ordnance and armoury.

The master-general and board order Captain Valentine Peyne, master-gunner of England, £30 for his extraordinary attention and pains in assisting in taking the general remains of all the stores within the office of his majesty's ordnance.

Jonas Moore, Esq. comptroller of his majesty's train of artillery will order into store the brass ordnance issued for the artillery company's public exercise at the artillery garden, viz.—

Brass ordnance on travelling carriages.

Two sakers, four 3-prs., one 8-in mortar.

Captain Valentine Peyne, master-gunner of England, is allowed to charge 1*s.* 6*d.* a day from the 9th August to 11th September for the gunners employed in mounting guns at Woolwich.

11 Nov. The Board direct there may be issued for the proof of ordnance at Woolwich in Mr Brown's yard, gun foundry, viz,—

Round shot for, with powder, match, &c.

12 cannon	8 demi-culverins
4 demi-cannon	16 sakers
23 culverins	1 3-pr.

Jonas Moore, surveyor-general, made Sir in this year.

1674. 2 May. His majesty's warrant to Sir Thomas Chicheley, kt. master-general, appointing him also master-general of the ordnance in Ireland, void by the death of Sir Robert Byron.

ditto. His majesty's warrant to Sir John Chicheley for the master of the ordnance and armoury, after his father, Sir Thomas Chicheley.

Edward Conyers appointed storekeeper to the ordnance.

Thomas Langley appointed second clerk to the master-general.

1676. 19 May. "Sir,

"His majesty having appointed a rendezvous of several of his majesty's horse and foot guards in Hyde Park on Tuesday next, being the 23rd of this instant, I desire you to cause eight field pieces, viz.
4 demi-culverins and 4 sakers brass ordnance, and two mortar pieces, with all their carriages, and furniture thereunto belonging, together with two wagons, two tumbrells and 4 tents attended with a competent number of gunners, fifty pioneers, with their respective officers in their best equipage, to attend the exercise of the said forces on the day above mentioned, and that they fail not to be three by eight o'clock in the morning at farthest.

"I am,

"Your most humble servant,

"MONMOUTH."

"To Sir Thomas Chicheley,
Master-general of his majesty's ordnance."

1677. George Legg appointed an assistant in the Board of ordnance with a salary of £300 per annum.

21 May. His Majesty's warrant granting Captain Richard Leake the master-gunnership of the Tower of London and Kingdom of England.

21 Nov. His majesty's warrant to John Brown, Esq. to be founder of ordnance, &c.

George Wharton, treasurer and paymaster, made Sir in this year.

1678. 1 May. His majesty's warrant appointing Colonel George Legg, general of the artillery, belonging to our army now, in the Spanish Netherlands, and to carefully discharge the trust of general of our artillery appertaining to the same, &c.

Charles Beaumont appointed keeper of small arms.

1679. The quarter books of the office of his majesty's ordnance, contain the allowance and wages to the several officers, clerks, artificers, labourers, and gunners daily, attendent and employed in the said office; also the wages, salaries and allowances of the several storekeepers, employed by the said office at Woolwich, Chatham, Sheerness, Upnor Castle, Windsor Castle, Portsmouth, Berwick and Hull; also to Captain Richard Leake, master-gunner of England, for exercise of scholars to shoot in great ordnance at the artillery ground.

To Captain Hugh Kelly for his attendance on the proof of ordnance and powder.

To Jonas Moore, Esq., for his better encouragement in the study of mathematics, especially in the art of fortification, gunnery, and artillery, and such other parts thereof as may most conduce to his majesty's service, as per warrant, 1st October, in the 24th year of his majesty's reign after the rate of £100 per annum.

To John Linger, firemaster as by warrant 11th Feb. 1675.

23 June. Sir John Chicheley, kt., Sir Wm. Hickman, kt., Sir Christopher Musgrave, kt., to perform the duties of master-general by commission.

1 July. George Legg appointed lieutenant of the ordnance, *vice* David Walter, Esq.

Thomas Langley appointed clerk to the master-general.

July Quarter Book. An allowance to George Legg, Esq., in consideration, that the mannor house in little minories, heretofore appropriated to the place of lieutenant-general of the ordnance is alleniated from the same place, without any benefit or satisfaction made him.

To Richard Wharton, gentleman, for his encouragement to travel into foreign parts, and perfect himself in the studies of mathematicks, especially in the art of fortification, and other parts thereof, as may render him capable to serve his majesty for the future as an engineer, as per warrant 1677.

To John Tinker, for his invention of a new way of shooting hand granades out of a small mortar piece, which may prove of great advantage to his majesty's service. Warrant dated 13th July, 1678.

To Sir Bernard de Gomme, kt., his majesty's chief engineer. Warrant dated 12 June, 1669.

Sept. Francis Smith, gentleman, Captain John Piggott, for attendance on the proofs of ordnance and powder.

13 Sept. Demand of Sir Bernard De Gomme for new carriages to be placed within the royal citadel of Plymouth.

31 ship carriages, 25 standing carriages, and the repairs of the old

carriages, in number 16, according to the note herein enclosed, and 7 carriages for St. Nicholas island near Plymouth.

New Fashioned Ship Carriages for Platforms.

	Carriages.		£	s.	d.	
Whole culverings drake	11	at	3	15	0	Not painted.
Demi-cannon drake	2	"	4	10	0	
Demi-culverings drake	5	"	3	0	0	
8-prs.	3	"	3	0	0	
6-prs.	5	"	2	18	0	
Sakers	2	"	2	18	0	
5-prs.	1	"	2	18	0	
Minions	2	"	1	18	4	
	31		101	5	8	Amount making not painted.

Standing Carriages.

	Carriage.		£	s.	d.	
Whole culverings	8	at	11	0	0	Not painted.
12-prs.	2	"	11	0	0	
Demi-culverings	9	"	9	0	0	
6 and 5-prs.	4	"	7	0	0	
Sakers	2	"	7	0	0	
	25		223	0	0	Amount making not painted.

Standing Carriages for St. Nicolas Island.

	Carriage.		£	s.	d.
Sakers of 10 ft. long	1	at	7	0	0
Sakers of 9 "	1	"	7	0	0
Demi-culverings of 9½ ft. long	1	"	9	0	0
" " 10½ "	1	"	9	0	0
" " 10 "	1	"	9	0	0
" " 9 "	1	"	9	0	0
Sakers of 8½ ft. long	1	"	7	0	0
	7		57	0	0

Abstract.

	£	s.	d.
Ship carriages	101	5	8
Standing carriages	233	0	0
" " St. Nicholas Island	57	0	0
Repairing old carriages	121	0	0
	£512	5	8

1682. 8 Jan. His majesty's warrant appointing George Legg, Lord Dartmouth, master-general of the ordnance.

ditto. Sir Christopher Musgrave appointed lieutenant of the ordnance.

His majesty's warrant (reciting that abuses had crept into the garrisons), puts all the master-gunners in Great Britain under the authority of the master-general, authorizing him to examine them, and turn out such as are unfit for service, notwithstanding they might have been appointed by patent from the exchequer.

17 Mar. Estimate by Sir Bernard De Gomme, his majesty's chief engineer, and assistant surveyor of the ordnance for dividing of the great house at Woolwich, &c., and repairing and altering the smith's shop there.

Apartments for the Storekeeper.

A parlour, kitchen, pantry, three rooms above stairs.

A hall common to the storekeeper, and Captain Leake, master-gunner of England.

Apartments for Captain Leake.

2 parlours, kitchen, pantry, 2 bedrooms.

For the lieutenant of his majesty's ordnance.

A great dining room, to be kept for the use of the officers of the ordnance, and in their absence Captain Leake shall have the use of the said dining room, and the closet; a great bed-chamber with a closet to be for the lieutenant of the ordnance; a little room for his servants; the third story for the servants of the storekeeper, and Captain Leake's family; a stable for Captain Leake's horses; a brewhouse to be common to the two families; fitting up part of the smith's house for 12 gunners; three rooms, each to contain four gunners.

2 June. An order for the saltpetre storehouse to be repaired.

1683
24 Jan. His majesty's warrant for building a storehouse[1] for the train of artillery, with the carriages and all other necessaries, and a new armoury for the use of the guards in St James's Park, on the south side of SpringGarden wall, according to a draught annexed and approved, for the so doing this shall be your warrant, &c.

CHARLES R.

1683,
25 July. Right trusty and well-beloved Councillor,

"We greet you well. Whereas we have taken into our princely consideration and care the great consequence and necessity of preserving the state of our artillery, munition, arms, and of all other the habiliaments and equipage belonging to our Magazine Royal within the Office of our Ordnance, which hath heretofore with much providence and wisdom of our predecessors been instituted, endow'd, and supported as a member highly importing the safety and defence of our dominions, as well by sea as land: We have therefore thought it requisite and necessary for the better conservation of our said office, and for the good of our service therein, that instructions should be forthwith drawn for the government of the Office of our Ordnance, under our Master-General thereof, expressing as well the particular and proper dutys of each principal officer severally,

[1] Now the "Gun House," St. James's Park.

as the common dutys of them jointly, in the general management of our said office, with the dutys likewise of each inferior officer, minister, and attendant thereto belonging; and forasmuch as you have, in obedience to our commands, presented to us a book containing the following rules and instructions for our royal allowance and establishment: We having taken the said instructions into our princely consideration, have thought fit to approve, ratify, and confirm the same under our royal sign manual upon every leaf of the said book, and do hereby declare our will and pleasure to be, that the said instructions so by us approved, ratified, confirmed, and signed, be and remain the established rules, orders, and instructions for the future government of the office, and officers of our ordnance under you the present Master-General thereof, and the Master or Masters-General of our ordnance successively. And our will and pleasure further is, and we do thereby strictly charge and require you, to govern yourself, and our office and officers under you, according to the literal sense, and tone, intent, and meaning of the said instructions, and if any of our principal officers shall misdemean themselves in the just execution of these instructions, either in their respective duties or in their common actings to our damage or disservice, or in neglect or disobedience to you in the discharge of their duty to us, then and in such cases you are to suspend any of the said principal officers so offending, and to acquaint us therewith, and the nature of the offence, that our will and pleasure may be further signified for removal or final deprivation of his or their employment, if we shall so see cause or think fit. And that there may not be wanting a fit and just support answerable to the dignity of the office of Master-General of our Ordnance, without having recourse to those undue means formerly practised, to the great disservice and unknown expence of our Treasurer, we have taken the same into our serious consideration, and have and do establish a plentifull and sufficient sallary out of the ordinary of our Office of the Ordnance, to be paid upon the quarter-books of the said office to you, or to the Master-General of our Ordnance for the time being, or any other executing the same place, to the end, that neither you nor any coming into or executing the said place after you, shall either by him, or themselves, or by any other person take or receive any sum or sums of money, presents, gratuities, or other consideration for any warrant, place, or employment whatever within the said office, the selling of which places and employments hath been and must necessarily (if continued) be the cause of many great mischiefs to our service; wherefore, we do strictly require the due and just performance of this our order from you, him, or them, or any that shall come after you in the same employment upon pain of our high displeasure; and whereas we have thought fit to regulate and establish the future annual allowance, not only to you, our Master-General of our Ordnance (or others executing that place), but likewise to all our principal officers and all other our inferior officers and ministers attending our service upon our quarter-books of our said office over and besides the fees payable to you or them upon their respective letters patents, out of our Exchequer, to the end our officers may for the time to come be encouraged to act faithfully and justly in their respective places; and you are not to suffer any increase to be made upon our quarter-books, either of places or additions to the places now established, without warrant under our privy signet and sign manual authorizing the so doing; and lastly, it is our further will and pleasure that all the sallaries, allowances, and wages hereby established and annexed be born and continued upon the quarter-books of our said office, but only during our royal pleasure. For all which this shall be to you, him, or them that shall come after you a full and sufficient warrant. Given at our Court at Whitehall, this five and twentieth day of July, 1683, in the five and thirtieth year of our reign.

By His Majesty's command,
L. JENKINS.

To our right trusty and well-beloved Councellor, George Lord Dartmouth, Master-General of our Ordnance, and to the Master or Masters-General of the Ordnance for the time being.

INSTRUCTIONS for the GOVERNMENT of our OFFICE of ORDNANCE under our MASTER GENERAL thereof, committed to five "Principal Officers"—our

LIEUT GENERAL OF THE ORDNANCE,
SURVEYOR GENERAL OF THE ORDNANCE,
CLERK OF THE ORDNANCE,
KEEPER OF THE STORES,
CLERK OF THE DELIVERIES,

Which said instructions we will and require be, by all and every one of them, faithfully and diligently observed and obeyed.

L. JENKINS.

C. R.

1683. *The proper duties of our* LIEUTENANT-GENERAL *are:—*

1. In absence of the Master of the Ordnance to receive all letters, instructions, orders, and warrants directed to our said office, and to impart the same forthwith to the rest of our principal officers, so as they may be duly executed and kept in the office.

2. To keep a minute-book or journal for his better information and cognizance of all transactions at the Board of our said office, as likewise an entry of all warrants, orders, instructions, and significations directed as aforesaid to himself and the rest of our principal officers.

3. By order of the Master of the Ordnance, with the advice and concurrence of our principal officers or major part of them, to consider of all estimates or demands for moneys, to be made for supply of our stores, furnishing our ships, forts, and castles, or carrying on any works or buildings in our fortifications, or any other our services, and to see (as

much as in him lyes) that both the said estimates and all bargains, contracts, and agreements thereupon drawn be such as may be for the advantage and good of our service.

4. To take care that the train of artillery in our Tower of London, and all its equipage, be fitted for motion upon any occasion (when it shall be ordered to be drawn into field for our service), and that all other marching trains in any of our magazines within our kingdom of England be kept in good condition, and ready for our service when required.

5. To direct and oversee the practice of the master-gunner, firemaster, mates, gunners, and fireworkers under them, and frequently to require an accompt of the master-gunner and firemaster of all our feed gunners, fireworkers, practitioners, and scholars of their proficiencies, and from time to time to acquaint our Master of our Ordnance therewith.

6. In absence of the Master of our Ordnance, to give directions in writing under his hand for the shooting off of great ordnance at our Tower of London upon the annual solemnity of our Coronation Day, and upon our or our Royal Consort's birthdays, reception of any foreign prince or embassadors, or upon any other extraordinary occasion of that nature whensoever appointed, which direction is to make express mention of the said appointment, and to specify the number and nature of the ordnance to be fired.

7. He is generally (as much as in him lies) to see that all officers, inferior ministers, artificers, &c., perform their respective duties, and to inform himself of their deportment and behaviour; and as any misdemeanors or good services shall come to his knowledge, he is forthwith to acquaint our Master-General therewith, to the end due punishment or reward may be given to every one according to their demerit.

The proper Duties of our SURVEYOR-GENERAL *now and for the time being are:*

1. To survey all stores and provisions of war belonging to our office, in the charge of our Storekeeper, who is hereby charged and required to give to him free ingress and egress at all convenient times.

2. To see the several species of stores distinguished and placed for their preservation as they ought to be, and in case they are not, to represent the same with their condition and qualities to the Master of our Ordnance, and Principal Officers at their next meeting.

3. To take care that all labourers, artificers, and workmen employed in our service be strictly kept to their labours and duties, and not to suffer any person (without cheque) to be absent, or loyter, or to make bad or slight work, or to misemploy, waste or embezle any stuff or stores.

4. To see that the person appointed clerk of the cheque do keep a particular account of the daily and weekly work, according to the rules prescribed him in his particular instructions.

5. With the assistance of our Principal Officers or the major

part of them, together with our master-gunner in cases where it shall be requisite, and our proof-masters, or one of them, to survey and make proof of all ordnance, powder, small arms, and all other emptions, and provisions of war, and not to suffer any stores to be received in, which are not good and serviceable, and also duly proved and marked with our mark,[1] if it ought so to be. And such stores and provisions, as upon survey and proof shall be found not to be agreeable to contract, agreement or warrant, nor fit for our service, immediately to cause to be taken back by the artificers or merchants that brought them in, that our magazines may not be unduly or unnecessarily charged therewith.

6. To procure and allow all bills demanding payment of money for goods delivered into our stores or for works and services done. And to allow only those under his hand that are agreeable to warrants, contracts or receipts, which allowance (to prevent unnecessary attendance and delays) is to be made by our said surveyor, within ten days after the receipt of any such bills, or to give his reasons to the Master of our Ordnance or our Principal Officers at their next meeting why he does not allow the same.

7. By himself or some of his sworn clerks (by him appointed) to assist at the taking of all accounts and remains, and to survey all stores of what nature or kind soever returned into our magazines from any of our ships, castles, forts, captains, gunners, or any other person or persons to whom the same were formerly issued, before our storekeeper be chargeable therewith, and to distinguish the good and serviceable from the repairable, and the repairable from the utterly unserviceable,[2] and to report to the Master of our Ordnance, or our principal officers at their next meeting, the particular defects of those that are to be repaired, that timely order may be taken for refitting the same, and making them useful for our service.

8. When any work, buildings or repairs in and about our fortifications,

---

[1] "Marked with our mark." Here we appear to have the first official reference to the well-known mark of the Board of Ordnance, viz., the Broad Arrow with the accompanying initials B. ↑ O. The question of the derivation of the name of this mark, viz., the Broad Arrow, has given rise to much controversy and ingenious research, but the origin commonly assigned to the "mark" is, I think, quite untenable. It is not necessary to go beyond the text of this warrant. "Our mark" was, of course, the Board's mark, and the mark or crest of the Board of Ordnance was the "Thunderbolt of Jove," as represented by the bundle of arrows in a clenched fist. The motto of the Board of Ordnance was "Suâ tela tonanti," where the word *tela*, originally meaning "arrows," included all projectiles. The "mark" used for the Stores of the Office of Ordnance would thus naturally be the crest of the Board of Ordnance, and although cannon might conveniently be marked with the whole crest (the Thunderbolt), such a mark would be too large for the smaller articles in charge of the Board, and a single barbed arrow-head would therefore be used as the representative of the more elaborate mark, or crest of the Board. Hence we conclude that the "Broad Arrow" is merely a corruption of the "Board's Arrow," which, when practicable, was always accompanied by the letters B. O.;—thus, B. ↑ O. The original mark has been adopted by the successors of the Board of Ordnance, and the letters W. D. have replaced the original B. O.

[2] This classification of stores as serviceable, unserviceable, and repairable still holds good,

or for other our services, are to be undertaken, to compute and calculate the charge thereof, and to propose the same to the Master-General of our Ordnance, or to our principal officers or major part of them, expressing the particular scantlings and dimentions of the materials therein to be used, and the lowest prices for which they may be had, likewise to survey all the works, buildings or repairs, taking the assistance of skilfull and experienced men if it be requisite, and carefully to examine that they be well and sufficiently done, whether undertaken by the great, daywork, or upon any other contract or agreement, and when finished to measure the same, and certify to the Master of our Ordnance, or Principal Officers, the just and due performance of the respective undertaking before any bills be allowed for debentures to be passed in order to the payment thereof.

9. To keep two counter or cheque books by way of journal and ledger of the receipts, issues, returns, and remains of all our stores of what nature soever chargeable upon, or accountable for, by our Storekeeper in the same method and manner as is by us enjoyned and prescribed to be kept by our Storekeeper.

The proper Duties of our Clerk of our Ordnance *now and for the time being:*

1. To keep a book of all orders and instructions given for the government of our Office, and to record all patents and grants, and the names of all officers, clerks, artificers, attendants, gunners, and others who enjoy the said grants, or any fees from us for the same.

2. To keep a journal book of the entrys of all warrants and orders for services to be done, provisions to be made, or supplies to be sent, and all directions and significations thereupon given; and to keep the originals thereof for the justifications and satisfaction of the Office, and for the particular charge and discharge of our Storekeepers upon their accounts, true copies whereof they our said Storekeepers are permitted to take.

3. To draw all estimates for provisions to be made, or supplies ordered to be sent, as aforesaid, to any of our ships, forts, castles, or garrisons, or for any other our services, to be undertaken in and by our said Office, and to register the said estimates and letters under our Privy Seal thereupon, and to keep entries of all warrants and orders from our Lord Treasurer, or Lords Commissioners of our Treasury, for moneys assigned or to be received for the use of our said Office.

4. To draw and indite all letters, instructions, commissions, deputations, contracts, or bargains, and other writings agreed upon by our Master-General and Principal Officers or our Ordnance, or the major part of them, for any of our services, and to keep thereof particular records.

5. Likewise to make out warrants to be signed by himself and our Surveyor, to the several artificers for bringing in the respective provisions ordered them to provide upon the several estimates and

contracts thereupon by them undertaken, and to enter the said warrants in a ledger book; likewise to draw bonds for performance of covenants.

6. To assist at the taking the remains of any of our ships, castles, forts, and garrisons, and to make certificates upon the accounts of the commanders, captains, and gunners thereof, to be signed by himself and the rest, or major part of our Principal Officers, first examining and comparing the said accounts by balancing the charge and expense thereof, computed both in specie and value.

7. To make debentures for all provisions and stores of war bought and brought into our magazines, and for freight, land and sea services, and other incident charges upon bills allowed of, by our Surveyor, pursuant to warrants, orders, and significations, either from us, our Royal Council, or of our Master-General of our Ordnance, or upon contracts and orders of the Board of our said Office grounded thereupon, and thereof to make enteries in his ledger book of debentures, likewise to draw all bills of imprests for monies ordered to be advanced upon any contract, agreement, or bargain made for our service, and to enter the same in his book of imprests ordered to be made.

8. To vacate all bills of imprests made for advance of monies upon any contracts, agreements, or undertakings as aforesaid, or other our service, by making a memorandum on the margine of the debenture to be passed, for satisfaction of the said contracts, agreements, undertakings, or other services when performed, that our Treasurer of the Ordnance may take notice how much is to be by him defaulted thereout, and recharged upon his account of cash, and if it shall appear upon the account of our Office that the neglect of our said Clerk of our Ordnance herein hath damnified us, that then and in such case he be answerable to us for the same.

9. That he or some of his sworn clerks attend at the receipt of all stores, and make indorsements upon the orders, bills, or warrants directing the bringing in of the same, by putting to his or their hands attesting the respective quantities received, and to enter the same in his journal of receipts, that so the Keepers of our Stores may be duly charged therewith.

10. That he or some of his sworn clerks do likewise attend at the return of all, or any of our stores formerly issued, and to examine the bills of lading or other writings that specify the returns so made, and the respective quantities thereof, and to note the times when, the persons from whom, and the places from whence returned, and to cause the same to be entered in the Journal of Returns and Remains, by him kept, that so our storekeepers may be recharged therewith.

11. To keep likewise annual leidgers of the receipts, issues, and returns of all our stores, as well at our Tower of London, as at any of our magazines or storehouses at Woolwich, Tilbury, Sheerness, Chatham, Portsmouth, Plymouth, Hull, or elsewhere, deduced from the journals or monthly abstracts of the respective storekeepers of our said magazines to serve as a cheque or controll of their accompt.

12. To keep a like annual ledger of all monies, tally's, and orders received by the Treasurer of our Ordnance, either out of our Exchequer upon imprests, certificates, or upon the sale of any sort of decay'd or

unserviceable stores or otherwise for our service, chargeable upon his voluntary account, as likewise of all monies by him thereout paid upon debentures, bills of imprests, or quarter books of our said office duly vouched and listed [1] to remain in our office for a controlment of our said Treasurer's accounts.

13. To make quarter books of the salaries, wages, and allowances due to our several officers, clerks, artificers, labourers, and gunners belonging to the office of our Ordnance, together with duplicates thereof (to be signed by all our Principal Officers, or the major part of them) for the Treasurer of our Ordnance to pay by.

14. To draw lists to be signed by the Master of our Ordnance for the payment of such quarter books and debentures as he shall judge necessary and convenient to be paid, together with duplicates thereof for our Treasurer to pay by, and to cause an affix of the lists so signed by our Master of our Ordnance to be set up at the door of our office, to give notice to the respective persons therein concerned to receive their monies ordered them thereupon.[1]

15. To draw quarterly a state of the debts of our said office, both ordinary and extraordinary, as well what part into debentures, as growing due upon bills and contracts, to be presented to the Master of our Ordnance for his better information, in order to the making due and timely provision for the discharge thereof, and this quarterly state of the debts so drawn is to be delivered to the Master within ten days at farthest after the determination of each quarter.

16. Lastly the said Clerk of our Ordnance is strictly commanded and required not to take any debenturage, untill it shall appear to us that it shall be convenient for our service to allow the same to be taken as formerly, nor to demand or receive any fees other than the ordinary antient and usual fees following, viz.:

For the entry of any patent, commission or warrant from us for constituting any officer or minister upon the quarter books of our said office	£2	0	0
For entering the warrants or significations for admission of any under minister, clerk, artificer, or tradesman ...	£1	0	0
For entering the commission for any gunner or warrant for any labourer or workman in the office of our Ordnance, or armory to be borne upon the quarter books...	£0	10	0

The proper Duties of our STOREKEEPER *now and for the time being are:*

1. To take care of all our ordnance, munition and stores belonging to the office of our Ordnance within our Tower of London committed to

[1] As no man could receive pay until he was listed or enrolled, so a recruit who did receive the pay of a soldier in accepting "the shilling" was held to be duly "listed" or "enlisted"—hence the origin of the term to enlist a recruit.

his charge and custody, and to indent and put in such ample and legal security for the safe keeping thereof, and for making true, and annual account for the same, as our Master of our Ordnance for the time being shall think reasonable, and the Lords Commissioners of our Treasury, or our Lord High Treasurer for the time being shall approve of.

2. To receive no stores whatsoever into our magazines so as to be charged therewith upon account, without warrant in writing signed by our Surveyor and Clerk of our Ordnance, grounded either upon signification from our Master of our Ordnance, or upon contract made by our Principal Officers, or major part of them, nor untill they have been first surveyed by our Surveyor, proved by our proof masters or one of them, and marked with our mark if it ought so to be, the attestation of which survey is to be indorsed upon the warrant or order that directs the bringing in of the said stores by our Surveyor, or one of his sworn clerks, signifying the serviceableness allowed, and quality thereof; our Clerk of our Ordnance or some of his sworn clerks at the same time making the same indorsement upon the said warrant, expressing the respective quantities so surveyed, proved, and to be received, and our storekeeper or one of his sworn clerks making the like indorsement thereupon, to attest what is by him received, and when such warrant is completed, or so much brought into our stores as is thereby directed, then our Storekeeper to give a bill of receipt of the total of stores thereupon brought in within ten days after the last delivery.

3. To issue or deliver none of our stores under his charge without a warrant or proportion in writing agreed upon by our Principal Officers, and signed under the hands of three, or more of them, grounded upon warrant under our Royal Signet and Sign Manual, or order of our Council or of our Lord High Admiral of England, or Lords Commissioners of our Admiralty for the time being for matters concerning our navy and sea affairs, together with the signification of the Master-General of our Ordnance for the time being, thereupon provided that such quantity of our powder as shall be thought meet to be spent in shooting off of great ordnance at our Tower of London, upon the annual solemnity of our Coronation Day, and upon ours, or our Royal Consort's birthdays, reception of any foreign prince, or embassadors, or upon other extraordinary occasions of that nature, whensoever appointed; or if powder and other stores to be repaired and returned, or for powder, shot, &c., that be for proving of great ordnance or small guns, or for any experiment to be therewith made, be delivered without any other warrant or order than from the Master-General of our Ordnance, or, in his absence, the Lieutenant-General thereof for the time being; and in case of sudden fires by a proportion, to be signed under the hands of our Principal Officers, or the major part of them present, which is to be made known and accounted for to our Master-General or the Principal Officers of our Ordnance at their next meeting.

4. To produce, or cause to be produced, monthly or oftener, at the Board of the Office, all such proportions as have been signed by our principal officers, for the issuing of any stores or provisions of war, to be compared at the board with the indents taken by our Clerk of the Deliveries, that so our Storekeeper may not have credit allowed him

for more than what shall appear to have been really and actually issued, as is hereafter more fully expressed in our instructions for the general and common duties of our officers.

5. To receive back into our storeshouses no provisions formerly issued to any officer, captain, gunner, workman, or other, without the same be first duly resurveyed by our Surveyor, or some of his sworn clerks, setting down the conditions and qualities, expressing the several defects, and distinguishing which of them are serviceable, which repairable, and which utterly unservicable; nor unless the Clerk of our Ordnance, or one of his sworn clerks, have notice and be present in order to the registering the said stores and provisions so returned in his book, entitled the Book of Remains received back into our Office, that hereby our said storekeeper may be again upon account.

6. To represent from time to time, to the Master-General of our Ordnance, or to our Principal Officers at the Board of our said Office, the several defects and wants of reparations and accommodations of our storehouses.

7. To keep a journal daily, and duly to record all receipts, issues or returns of our provisions, and all the wants, losses, lendings and embezlements whatsoever.

8. To make an annual account, deduced into ledgers from the said journal, of all receipts, issues, and returns of all our stores and provisions of war, under his charge, which said accounts or ledgers are to be in the method and manner following, viz.:—

9. The first account of our present Storekeeper to commence from the remains of our said stores and provisions of war taken upon, or next after his entrance upon the said office, as the basis and foundation of all his subsequent accounts, and to terminate at midsummer following, and so to continue de anno in annum, until the last general remain, anno 1679, and that then he compare his account therewith, and to adjust the true state of both, by balancing the several species thereof according to their qualities.

10. That these accounts, and all others for the future which are to comprehend the receipts, issues and returns of all our stores and provisions of war, chargeable upon our present Storekeeper, or Storekeeper for the time being, be kept in two large books containing all the several species and natures of stores, in columns distinct and fairly written.

11. That they be kept by way of debtor and creditor, and balanced throughout, more especially every year at midsummer, and delivered into the Board of the Office with sufficient vouchers at Michaelmas following, or within fourteen days next after, to be examined, and subscribed by our Principal Officers, as is enjoyn'd by the General Instructions, after which examination and subscription immediately to deliver the same to the auditors of the imprest for passing his account.

12. That one of the books do (on the left hand page thereof) exhibit and make him debitor for all manner of serviceable provisions and stores, whether received (de novo) into our magazine, or otherwise, upon any contract, or whether returned from sea, or elsewhere, in a

serviceable condition, according to their nature, number, weight and measure, when and from whom received, upon what contract or order, and at what rates.

13. That on the other page thereof he have credit given him for all manner of such serviceable provisions and stores issued, expressing the nature, number, weight and measures thereof, when and to whom delivered, by what warrant or order, with the respective values thereof.

14. That the other of these books do in like manner (on the left hand page thereof) exhibit and make him debitor for all manner of provisions and stores returned from sea, or elsewhere, which, upon a due survey or proof thereof made, shall be found to be in any wise impaired, defective, decay'd or unserviceable, and so not thought fit to be mixed with good and serviceable, and this also in nature, number, weight and measure, when and from whence returned, with the prime value thereof.

15. That on the other page of the same book he have credit likewise given him for whatsoever thereof shall be thought fit to be issued to be repaired, or otherwise disposed of for our service, in number, nature, weight and measure, when and to whom issued, or how otherwise disposed of, by what warrant or order, with the prime value thereof. By which method the Office may at pleasure be readily satisfied what is or ought to be the state of any species under his charge, what hath thereout been issued, and for what services, all distinguished as aforesaid with the value of each respectively, and of the whole.

Which being added to the receipts and issues of all other provisions thus monthly accounted for by the other subordinate storekeepers will show—

1st. The true state and value of our stores in every one of our magazines distinctly and altogether.

2nd. The total of the annual expense.

Proper Duties of our Clerk of the Deliveries *in the Office of our Ordnance now and for the time being:*

1. To draw all proportions for delivery of any of our stores of war in pursuance of any warrant under our Royal Signet and Sign Manual, or order of our Privy Council, or of our Lordship Admiral, or Lords Commissioners of our Admiralty, together with the Master of our Ordnance, his signification, or, in his absence, the order of the Office Board thereupon; and to keep two books, one of the copies of all warrants, and orders, for the proportion, or delivery; the other a journal of the proportions themselves, or deliveries, of which he is to take and keep the indentures and other securities of the deliveries vouched under the hands of the respective receivers or accomptants, whether governors, colonels, captains, gunners, or any other persons thereunto appointed, whereby a true charge may be given to any receiver, or accomptant, when thereunto called or required; and the said journal he is to deliver to the Clerk of our Ordnance, to be by him,

or one of his sworn clerks, posted into a ledger, as in his instructions he is particularly directed.

2. Himself or one of his sworn clerks to be present at all deliveries of stores, and to assist in taking of remains, and upon the next proportion or supply drawing to, charge the remains with the new supply upon the respective ship, fort, castle or garrison, or upon the particular person into whose custody they shall be received.

3. To produce monthly, or oftener, at the Board of the Office the indents by him taken for any deliveries of stores, to be compared with our Storekeeper's proportions, and to serve as a cheque upon his accounts.

Proper Duties of our TREASURER OF OUR ORDNANCE *now and for the time being:*

1. At his first entrance upon the execution of his place to give such competent security to us for the true and faithfull discharge of his trust as the Master-General of our Ordnance now or for the time being shall judge necessary, and by the Lord High Treasurer of England or Lord Commissioner of our Treasury for the time being shall be approved of.

2. To solicit the Lord High Treasurer of England, or Lord Commissioner of our Treasury for the time being, for moneys upon all estimates and demands presented to us, our Privy Council, or their Lordships, and to procure our Letters of Privy Seal, and all other necessary warrants and orders for the same, and thereupon to receive all monies so assigned and directed for our service in the Office of our Ordnance, and to acquaint our Master-General and Principal Officers with the particular sums and times of receipts.

3. To pay all monies upon quarter books, orderly debentures, or bills of imprest, signed by three or more of our Principal Officers (whereof our Clerk of our Ordnance to be always one) and listed by our Master-General, and to take the hand or acquittance of the parties concerned to acknowledge the payment thereof.

4. To take no poundage, fee, gratuity, or reward, directly or indirectly, from any person or persons for or in consideration of any of our treasure to be by him paid upon quarter books, bills of imprest, debentures or otherwise, and if he shall be found guilty thereof by our Principal Officers, that then, upon their representation of the same to the Master-General of our Ordnance, he be suspended from his employment untill our Royal Pleasure be thereupon declared.

5. To present weekly, or oftener if required, to our Master-General of our Ordnance, or in his absence to our Lieutenant-General and the rest of our Principal Officers, a state of our cash, in writing, subscribed by him of all monies, tallies, and orders of our exchequer or received for our service weekly since the first state of our cash, charged upon him by our Letters of Privy Seal, bearing date the and

what thereof hath been paid or assigned weekly, and for what particular services, that so the balance of the said state may appear strictly to agree with the monies, tallies, and orders remaining in his hands from time to time.

6. To pay no money, or make assignment of any order, or tallies for the service of our said Office upon quarter books or debentures, untill he be directed so to do by list or lists, signed by our Master-General of our Ordnance, or in his absence by three of the Principal Officers at least, whereof the Lieutenant-General to be always one, therein expressing the quarter book, the date and sum of each debenture, and the persons to whom payable, the same to be paid in course according as they are listed.

7. Not to vacate or cancel any bills of imprests for money advanced, or to be advanced for any of our services whatsoever, but upon the proper debentures upon which they ought to be vacated or cancelled, and as he shall find directed in the margin of the debentures, and that all bills of imprests so sett off upon debentures, and vacated as aforesaid, be by the said Treasurer or persons therein concerned canceled and returned to the Clerk of our Ordnance next office day, that so both bill of imprest and debenture may not be continued in our Treasurer's hands, or be brought by him to accompt for one and the same thing.

8. To keep all monies by him received or to be received upon accompt of our service in the Office of our Ordnance in the house allotted him within our Tower of London, commonly known by the name of the Treasury or Pay-house, that the same may be ready to be viewed or inspected as often as the Master of our Ordnance shall think fit to direct, and that no monies be by him paid elsewhere but in the said Pay-house (without particular direction from the Master of our Ordnance, or in his absence by the major part of the Principal Officers), nor any monies whatsoever upon score of any payments to be by him diverted from the proper uses and services for which they were assigned without signification of the Master of our Ordnance.

9. Not to take up any monies at interest without particular warrant from our Lord High Treasurer or Lords Commissioners of our Treasury for the time being, grounded upon our letters of privy seal and signification of the Master of our Ordnance thereupon to the Board of our said Office, therein expressing the sum to be borrowed, upon what security, and what interest or other consideration shall be therefore allowed, and as soon as any monies so borrowed shall come to his hands to give notice thereof to our Master-General or Principal Officers of the person or persons' names who advanced and lent the same, the time when, and at what rates, that so entry thereof be made by the Clerk of our Ordnance, and when the principal or interest thereof, or both, be demanded, or come to be paid, the same may be cast up, and adjusted publickly at the Board of our Office, and allowed by the Master of our Ordnance, by direction of our Lord Treasurer or Lords Commissioners of our Treasury for the time being as aforesaid.

10. At Michaelmas every year, or within fourteen days next after, to deliver or cause to be delivered at the Board of our Office, an account not only of all monies by him received out of our exchequer within the

year, ending the last of June preceding, attested by one or more imprest certificates, under the hand of the Auditor of our Exchequer for the time being, but also of all other monies which have come to his hands within the said time, upon the sale of any sort of decayed stores, or otherwise for our service as the charge of his account.

11. At the same time, in order to his discharge, to deliver or cause to be delivered at the Board of our said Office an accompt not only of all monies by him paid within the said time upon debentures or quarter books, duly vouched and listed for payment by the Master of our Ordnance, or in his absence by the Lieutenant-General, and major part of our Principal Officers for the time being, but also of all monies by him issued and paid upon bills of imprest within the said time, and then standing in force undischarged, with the time when the persons to whom, and the service for which the same were impressed, to the end the said accompts may be examined at the Board of our Office, and returned to our said Treasurer, vouched and attested under the hands of three, or more of our Principal Officers, whereof our Clerk of our Ordnance for the time being (whose proper duty it is to keep a cheque both upon our said Treasurer and Storekeeper) to be always one, within twenty days next after such delivery thereof by our Treasurer at the Board of our Office, so that the said accompts may be delivered by him our said Treasurer, to one, or both of the Auditors of our Imprest for the time being, within six weeks at farthest next after Michaelmas every year to pass in due form.

12. That at the same time, viz., at Michaelmas every year, or within fourteen days next after, a duplicate of the said Treasurer's accompts be likewise delivered by him at the Board of our said Office to be likewise examined, and signed by three or more of our Principal Officers of our Ordnance as aforesaid, and there kept by the Clerk of our Ordnance for satisfaction of our Master-General and other the Principal Officers thereof upon occasion.

General Instructions for the common duties of our PRINCIPAL OFFICERS OF OUR ORDNANCE, *to be by them observed.*

Officers' abode.

To make their ordinary habitations and abode in the houses or lodgings assigned them in or near the Tower.

Their meeting.

To meet in the Office within the Tower of London twice a week at the least, viz., every Tuesday and Thursday about eight of the clock in the forenoon, or oftener as our service shall require, and there to consult together, and order all business appertaining to them concerning our service, or as often as the Master of our Ordnance shall direct, or our service require.

Obedience to orders, &c.

To observe, follow, and see performed all orders, warrants, or significations of the Master of our Ordnance, whether grounded upon warrants from Us, our Privy Council, or Lord High Admiral of England (for sea affairs), and in all things to obey him (jointly or severally in their respective places) as Master-General of our Ordnance, and such directions as from time to time he shall judge fit and necessary to give for our service.

Form of accompt to be observed.

To see that the accompts of our office, as well for monies, as stores, as likewise the cheque controll books thereof, be kept in the due form as by Us is prescribed in the proper instructions of our Treasurer and Storekeeper, and to take care that they be reasonably exhibited yearly to the auditors of our imprest as enjoined, and such of our Principal Officers whom it more particularly and respectively concerns to keep the cheque and controll books upon the said two accomptants are hereby required to be careful in the punctual discharge of their duties.

Supply of stores needful: how to be proportioned and estimated.

Upon the expenditure of our stores, and lessening of our magazine by occasion of any of our services, to assemble and meet together and consider of the want thereof, and accordingly to cause a proportion of what shall be needfull to be made, the same to be drawn up by our Clerk of the Deliveries, to be estimated by the Clerk of our Ordnance expressing therein the particular prices, as near as can be done, for which they may be provided, and the said proportion so estimated, signed under their hands, or the major part of them to deliver to the Master-General of our Ordnance, to be by him presented to us, our Privy Council or Lord High Treasurer, or Lord Commissioner of our Treasury for the time being, that monies may be ordered, and directed to be issued to the Treasurer of our Ordnance for carrying on of our services by resupplying our magazines from time to time with good and serviceable stores.

Stores: how to be provided.

Upon due order so given for authorizing such proportion and estimate, our said Principal Officers or the major part of them are to call before them such persons as they shall know to have any of the said provisions, and not to suffer the serving in thereof, to be ingrosed by any man, to the prejudice of our service, upon pretence of any office or grant intended for the advancement of the same, but to bargain freely, with the ablest and fittest artificers or merchants, and such as will be ready and willing to serve us in every kind the best stuff, and the best cheap,* which agreement and bargaining is to be made in the Office, by the major part of our Principal Officers, and entered and recorded by our Clerk of the Ordnance.

* "Best cheap" answers to the French meilleur marché." "Ceap" was a market.

Provisions and stores: how to be proved and received.

Before any such provisions agreed and bargained for as aforesaid be brought into our stores, our said Principal Officers or the major part of them (whereof the Surveyor to be always one), with the assistance of our proof masters, and of such others as the officers do know to be best able, and to view and try the same, whether they be good and serviceable, and being so found they are to cause the same to be brought into our stores, and delivered in their presence, or their sworn clerk's, upon such days and times as the provisions shall be appointed to be received by the Board of our said Office, or the major part of them, and not otherwise.

Provisions and stores: how to be entered and charged.

And that the receipts of our said provisions or any of our stores of what kind soever delivered into our magazines may be duly charged, we will, and require that particular entry thereof be made by our Surveyor, Clerk of our Ordnance, and Keeper of our Stores in their several transcripts of the journal books, expressing the particular kinds or sorts, numbers, weights, and measures of the said provisions and stores, and prices agreed upon, the names of the delivery and times when they were received.

Books: how to be signed and kept.

Which three books being found precisely to agree shall be subscribed by all the Officers (or the major part of them, whereof the Surveyor and Clerk of the Ordnance to be always two), and that book kept by our Storekeeper so subscribed to be delivered to the Auditors of the Imprests, whereupon the accounts may pass regularly and in due form, and that book kept by the Clerk of our Ordnance always to remain in the Office for the view, examination, and use of the Officers according to their respective places and duties.

Receipts and payments of money: how to be charged and accompted for.

The receipts and payments of all monies for our service are to be charged and accompted for in the method and manner by us prescribed in the particular instructions of our Treasurer, in the ledger book to be by him kept, and our Surveyor and Clerk of our Ordnance are in like manner to keep two cheque or comptrol books, which three books being upon examination at the board of our office found precisely to agree, that kept by the Clerk of our Ordnance shall always remain in our said office to be compared upon occasion with other two books kept by our Surveyor and Treasurer, and to be all three subscribed by our Principal Officers, or the major part of them, and that book kept by our Treasurer, and subscribed as aforesaid, is to be delivered to the Auditors of our Imprests yearly at Michaelmas, that so the accounts may pass before them in due form.

Debentures to the creditors: how and when to be made.

And that such satisfaction may be given to the several creditors as may assure them of due payment, our said Officers are to take care that within ten days after the full deliveries of any provisions or stores brought in by any artificers, or merchants, upon warrant, contract or agreement, a bill or bills of the provisions or stores delivered be made out and signed by our Storekeeper, acknowledging the receipt thereof; and from thence within ten days more to be allowed by our Surveyor, that so debentures may be made out and passed within one month's space after the final delivery of such provisions and stores by the Clerk of our Ordnance, to be signed by himself and the major part of our Principal Officers, vouchers thereunto specifying the numbers, weights, measures, and prices agreed upon.

Works and workmen: how to be ordered and accompted for.

For the better regulation of the daily working of artificers, workmen, and labourers, about the necessary works of our said Office, whether by the day or by the great, we will and command that the same be let by consent of all or the major part of our Principal Officers, and in such sort as may be most advantageous for our service, to which end the person appointed to be Clerk of the Cheque over the said artificers, workmen, and labourers, is by them to be called upon for a particular accompt of the daily and weekly works, which he is to keep in two distinct accompts, the one to be delivered weekly to our Surveyor, the other to the Clerk of our Ordnance, to remain in our Office, and to be signed and subscribed weekly by the major part of our Principal Officers, making particular mention both of all the artificers, workmen, or labourers hired and employed, and of all material or provisions as shall by any of them be brought in, expended or made use of daily or weekly in the works of our office, that so the bills to be brought in at the finishing of the services undertaken to be performed, may thereby be examined before they be allowed by our Surveyor, and no wages entered for longer time than the artificers or workmen have wrought, or served, and no materials or provisions admitted of or paid for, but such only as shall appear by the said cheque accompt to have been really employed or used.

All provisions to be

They are to take especial care that none of our provisions, of what kind soever, be used or employed by themselves to their own uses, or to the use of any other person, but be wholly kept and used for our immediate service; they are likewise to take especial care that no ammunition, or stores of what kind or nature soever be delivered, lent, sold or otherwise issued to any person or persons, out of our magazines without particular warrant in writing under employed in our service only.

our Sign Manual, or order of our Council, or letter of our Lord High Admiral, or Lords Commissioners of our Admiralty for the time being (if it concerns sea affairs), or signification from the Master-General of our Ordnance thereupon, and further to see indents between the parties to whom the said ammunition, provisions or stores are to be delivered, lent, sold or otherwise issued, or other caution to be taken to testify the delivery, sale, loan, or issue thereof, whereby such parties as receive the same may be called to answer.

Allowance for waste of stores.

They are to make reasonable allowance to our storekeeper for waste, or leakage, of all wasting, leakable and perishing stores from time to time upon view and inspection taken by the major part of them, whereof our Surveyor to be always one, which allowance is to be certified in writing under the hands of our said Principal Officers or major part, in order to our said Storekeeper's discharge.

Remains: how to be taken.

They are (as often as shall be requisite) to take the remains of any of our ships, forts, castles, or garrisons, and to examine the accompts of the respective commanders, captains, and gunners, and if upon examination of the state of the said accompt there shall be anything found amiss or contrary to the rules and instructions by us given and authorized, for regulating the expense of our munition and stores, to represent the same to the Master of our Ordnance, that order might be taken for the punishment or reformation of such errors, either by his own authority, or the Lord High Admiral respectively, or by the Lords of our Council, or Court of our Exchequer, or otherwise, as shall be most fit and requisite.

Manner of our officers acting jointly.

By the allowance, direction and signification of the Master of our Ordnance in all matters of weight, and in all courses by common advice and consent of most voices, and especially by following the literal prescripts of their warrants, to do and perform all the necessary services of our said Office within their respective charges.

To see all officers and ministers perform their duty, &c.

To prescribe to every one whom they employ their several charges and duties, and to see the due performance thereof, and to govern all under officers, and ministers, encouraging the faithful and punishing the negligent and unfaithful, by defalcation of wages, suspension from their places, and by moving the Master of the Ordnance for their further punishment or final deprivation as cause shall require.

No stranger or vagrant to be admitted into the storehouses.

To suffer no vagrant or suspicious person, or any foreigner or stranger (without knowledge of his quality, or some trusty person to attend them) to haunt or to have intercourse in the offices or storehouses, especially in the powder room, which, for more assurance, we will and command be kept under two locks with divers keys, whereof our Storekeeper may have one, and the rest of our Principal Officers the other, to be kept in their common chest, in the Office, whereof every one to have a key, so as there may be no access to the powder without the personal presence of two of them at least.

Allowances, sallaries and wages upon the quarter books: how settled.

And whereas we are given to understand that it hath been usual and customable in our said Office of the Ordnance, if any officer, clerk, or minister die within the current quarter, before the same be finished, that the allowance, salary, or wages due upon the said quarter when entirely finished, hath usually and of course been paid wholly to the executors, administrators or assigns of the officer, clerk, or minister deceasing within the said quarter, and not to the person succeeding, and hath been so accompted for, both by the Lieutenants and Treasurers of our Ordnance according to the practice of our said Office: Upon due consideration whereof we do hereby ratify, authorize, and confirm the said practice and custom. And this declaration of our royal will and pleasure for authorizing the same shall be as well to the Master of our Ordnance, the Principal Officers, and Treasurer of our said Office, as to the Auditors of our Imprest, and all other of our officers whom it may concern, sufficient warrant for allowing, paying, and passing thereof to accompt.

No gratuities or fees to be paid.

Our further will and pleasure is that no gratuity or other reward whatsoever shall be given or received by any officer, clerk, or minister belonging to our Office of Ordnance by reason of any bill, debenture, contracts, bargains, or payment of monies thereupon, or any other consideration or pretence whatsoever.

No clerks' places to be sold.

And whereas the selling of the clerks' places, under our Principal Officers hath been and must necessarily (if continued) be the cause of many great mischiefs to our service, we strictly command and declare that no Principal Officer hereafter presume to sell any clerk's place, and that the clerks of our several Principal Officers being of great trust, to the end that they may not be overburthened or discouraged in their duty, we require that every Officer keep the full number of clerks to their respective places as is directed and allowed by our establishment or warrants, and that no inferior servant, or unfit person be charged or entered upon our Office as a clerk, or that any of them receive less wages than they ought to have, and that upon breach of this order being made known to the Master of our Ordnance, that he immediately suspend the Officer so offending and acquaint us therewith, in order to the dismissing him of our service for the future.

Enrollment of these instructions to be made and oath to be taken by every new officer.

Lastly, we will, that these our instructions be forthwith inrolled and recorded with our Remembrancer of our Court of Exchequer, and that upon the admittance of any new Officer or Officers into our said Office, upon the death or removal of any of our present Officers, these our instructions be openly and publickly read at the Board of our said Office, and that such new Officer or Officers, before he or they enter upon the execution of his or their places, do make oath in our Court of Exchequer, before our Lord Chief Barron, and other our Barrons there, that he, or they, and every of them, will faithfully and justly discharge their respective duties in all things committed to their charge to the best of their industry, skill, and knowledge, in such manner as we have in these our instructions, for the better government of our said Office, particularly directed.

Instructions for the Secretary or Chief Clerk to the Master of our Ordnance.

His duty is to attend the Master-General of our Ordnance and to observe his directions, and follow such business in the Office as shall be given him by the appointment of our said Master-General, in order to our service, and to render an account to him from time to time of such proceedings as may relate to him in the execution of his charge.

To draw all significations upon any warrants or orders directed to the Master of our Ordnance, for our service, to be by him signed and transmitted to the Lieutenant-General of our Ordnance and the rest of our Principal Officers, and to make all commissions and warrants, to be signed by the said Master of our Ordnance, for the admission and entertainment of all artificers, tradesmen, labourers, gunners, or others to be employed in our service within our said office.

He is not to take any fee, gratuity, or other reward whatsoever over and above his sallary, and the sallary of his under clerk, or clerk allowed him, nor above forty shillings for a warrant for a labourer's place, nor above twenty shillings for a gunner's commission, nor above ten shillings for any other warrant or signification whatsoever.

To sit with and be present at the meeting of the Principal Officers in the Office upon office days, and upon all occasions of their general meeting, that he may be able to acquaint the Master of our Ordnance with all transactions at the Board, to the end he may be duly informed of the execution of all such orders and directions as shall from time to time be given for our service.

To make due entrys of all warrants, orders, or directions from us, our Privy Council, Lord High Admiral of England for Sea Affairs, and the Master of our Ordnance for the time being, and after entries so made, to deliver over such warrants and orders, &c., to the Clerk of our Ordnance, there to be registered and remain for the justification of our Master and Principal Officers and for Precedents, and further directions for the future government of our Office.

INSTRUCTIONS for the INFERIOR OFFICERS under the CHIEF OFFICERS OF OUR ORDNANCE, which are under ministers and attendants or tradesmen and artificers.

THE UNDER MINISTERS

Either attend the Officers as their clerks, or the office as the Deputy Keeper of the Armory, Keeper of the Small Gun Office, Storekeepers, Engineers, Master Gunner, Master Firemaster, Proof Masters, Waggon Master, Clerk of the Cheque, Purveyor, Messenger, Labourers, Gunners, &c.

THE CLERKS' SPECIAL DUTIES.

To attend the Officers to whom they respectively belong, and to write and keep their books fairly, and with diligence and truth, and to obey the orders and directions of the Master-General of our Ordnance or the Principal Officers in the general execution of the office.

Upon their admittance into their employments to take the oaths of allegiance and supremacy and all other oaths and tests by law required, and to give security, if demanded, for the faithful discharge of their duties.

To take no gratifications or rewards, over and above their sallaries, besides what may be made them by the Master-General of our Ordnance upon occasion of their pains and care taken in dispatch of any extraordinary service.

To give dispatch under their respective officers to all tradesmen, artificers, and others dealing with the Office, by making out duly, and in course, their warrants, contracts, bills, or debentures ordered to be made and signed, and if they shall be found guilty of giving them any wilfull or unnecessary delay to be punished answerable to their demerit, or dismissed from their employment, if the Master of our Ordnance shall so think fit to direct.

We likewise will and command that no clerk to any of our Principal Officers presume to take any additional clerks, or any person to write extraordinarily under them, so as to bring a greater charge upon the ordinary of our said office, without signification from the Master of our Ordnance or order of the Board, mentioning the extraordinary occasion and necessary service which is to be performed in due and convenient time, and that after the service is over the said extraordinary clerk or clerks are immediately to be dismissed, with such gratuity as the

Master of our Ordnance or the Board shall see reasonable, that no increase or charge may be brought upon us, or continued longer than is absolutely necessary for our service.

Proper Duties of the DEPUTY-KEEPER OF OUR ARMOURY, *now and for the time being.*

To take into his care and charge all our defensive arms and armour, as well for foot as horse, within our Armoury in our Tower of London, with all the necessary equipment thereunto belonging, and to enter into such legal security to us, our heirs, and successors for the safe keeping and rendering a just and true account from time to time as our Master of our Ordnance for the time being shall judge reasonable.

To see that all the arms and armour received well-conditioned into his charge be so kept, cleans'd, and oiled, as they may be fit for our service when required.

To take care that the armourers and labourers daily employed in our said armoury give their due attendance upon the necessary works enjoyned them, and not to suffer them to be absent, or loiter, or embezle any of our stores within his custody, and if any of them shall be found faulty to represent the same forthwith to the Master-General of our Ordnance or our Principal Officers, that course may be taken either for redress or punishment.

To receive no armour or any equipage thereto belonging into his charge before the same have been surveyed by our Surveyor, proved by our proof masters, and marked with our mark if it ought so to be, nor untill notice be taken by our Clerk of our Ordnance or some of his sworn clerks of their respective numbers, pursuant to contracts or warrants thereupon made, so that he may be duly charged.

To issue no armours, or any equipage thereunto belonging, without a proportion signed by three or more of our Principal Officers of the Ordnance, grounded either upon our warrant under our Royal Signet and Sign Manual, or order of the Lords of our Council, with the signification of the Master of our Ordnance, or in his absence the order of our Office Board thereupon, nor without an indenture thereof taken by the Clerk of our Deliveries for his discharge.

To keep journals and ledgers of all receipts, returns, and issues of our armours, and of all the necessary equipages thereunto belonging under his charge and keeping, and yearly to render an account thereof to the Master of our Ordnance and our Principal Officers at the Board of our said Office, to be there examined and signed by three or more of our said Principal Officers, and exhibited at Michaelmas every year to the Auditors of our Imprest.

Not to make use of, or cause to be expended any old defective armours, or pieces of armours, upon pretence of repairing others, until a survey of the said old armour be made by our Surveyor, and of the particular repairs necessary to be undertaken, and by him certified to the Board of the said Office, that order may be thereupon given, and a proportion for the delivery drawn and signed by three or more of our Principal Officers, and indenture taken by our Clerk of the Deliveries if it shall be needful.

Proper Duties of the Keeper of our Small Arms.

Upon the entrance into his place, to enter into such sufficient security to us, our heirs, and successors for the true and faithful discharge of his trust, as the Master of our Ordnance for the time being shall judge necessary.

To keep all the small guns and their equipage committed to his charge, in due order, and so disposed as may be best for a ready accompt and decent view, and that he cause (with the help and assistance of the furbusher) all such pieces as he receiveth well conditioned into his custody to be oyled and kept clean, or to undergo the charge of the damage we shall sustain by the detriment that may accrue to such pieces by his neglect of the duty enjoyned.

To receive no small guns or any other provisions or equipage thereunto belonging, into his charge before the same have been surveyed by our Surveyor, proved by our Proof Masters, and marked with our mark, nor untill notice be taken by our Clerk of our Ordnance or some of his sworn clerks of their respective numbers, pursuant to orders, contracts, or dividends thereupon made, that so he may be duly charged.

To issue no small guns, or other provisions under his charge, without a proportion, signed by three or more of our Principal Officers, grounded upon our warrant, or Order of our Council, or Lord Admiral, or Lords Commissioners of our Admiralty for the time being, with the Master of our Ordnance his signification, or in his absence the order of our Office Board thereupon, nor without an indentnre thereof taken by the Clerk of our Deliveries for his discharge.

To keep journals and ledgers of all receipts, returns, and issues of small arms, and all other equipage thereunto belonging, under his charge and keeping, and yearly to render an accompt thereof to the Master-General of our Ordnance and our Principal Officers at the Board of our said Office, to be there examined and signed by three or more of our said Principal Officers, and exhibited at Michaelmas every year to the Auditors of our Imprests, in the same method and manner as prescribed to the Keeper of our Great Stores.

Instructions for the Storekeepers of the Out-ports and Magazines.

To take into their respective charge and custody all such guns, gunners' stores, powder, match, shot, and other habiliaments of war whatsoever as shall be either sent from the Office of the Ordnance for a supply of our said magazines or returned from any of our ships, forts, castles, or garrisons, upon any remains taken by direction of the Master-General or Principal Officers of our Ordnance, and the same safely to keep, and to indent and give in sufficient security for so doing, and likewise for rendering a true and just accompt from time to time of all receipts, issues, and returns of stores under their several charges in the method and form hereafter required.

To send up to the Office of our Ordnance monthly abstracts of the daily receipts, returns, and issues that shall be made of all manner of stores, and in the last column of the said abstracts of issues to set down the remaining state of the stores at the month's end for the satisfaction of the Master-General of our Ordnance, and the Board of our said Office in the examination of the leidgers to be by them kept and sent up to our said Office every three months.

To keep three journals, one for receipts, one for returns, and one for issues, and a ledger wherein to post the said receipts, issues, and returns from the said journals, which ledger of stores, duly and fully posted, cast up and stated, distinguishing the good and serviceable from the repairable and unserviceable, with the days of the month when all such stores were received and returned, and from whom, as also the days of the month when such stores were issued, and to whom, and by what warrant. They are every one of them respectively charged to send up at the end of every quarter of a year to the Board of our said Office to be compared with the monthly abstracts of their journals, for the better clearing of their accompts.

Not to deliver out or issue any of our stores of what nature soever under their charge without a proportion signed by three of our Principal Officers at the least, grounded either upon warrant from us, our Privy Council, or letter of our Lord Admiral, or Lords Commissioners of our Admiralty for the time being, and the Master-General of our Ordnance's signification thereupon.

To take care from time to time to give the Right Honourable the Master-General of our Ordnance not only an accompt of the condition of our stores under their respective charge, of their wast and decays, but also of the condition of the storehouses wherein the said stores shall be lodged, and not to put any workmen to work upon the repairs of any of our said storehouses without first giving a particular account of the several defects of the said storehouses, and an estimate of the charge, at the cheapest rates that may be, of repairing and making the same good, and thereupon to receive the order of the Master-General of our Ordnance for putting workmen to work upon the same, as he shall think fit to direct, and to pursue such orders, and not otherwise.

Lastly, they are not to hire any extraordinary labourers, nor to lay out any monies upon accompt of the office, without first giving notice to the Master of our Ordnance of the occasion or service, and upon what order and directions they shall then receive from the said Master of our Ordnance they are to proceed accordingly, and not otherwise; in which laying out of monies as there shall be occasion, you are monthly also to send up to the Master of our Ordnance a particular accompt what they shall so lay out, without the continual practice of which method, no allowance shall be made them of any monies they lay out.

Instructions for our PRINCIPAL ENGINEER.

He ought to be well skilled in all the parts of the mathematicks, more particularly in stereometry, altimetry, and geodasia, to take distances, heights, depths, surveys of land, measures of solid bodies, and to cut any part of ground to a proportion given, to be well skilled in all manner of foundations, in the scantlings of all timber and stone, and of their several natures, and to be perfect in architecture, civil and military, and to have always by him the description, or models, of all manner of engines useful in fortifications, or sieges, to draw and design the situation of any place, in their due prospects, uprights, and perspective, to know exactly the rates of all materials for building of fortifications, thereby to judge of any estimates proposed to him to examine.

To keep perfect draughts of every the fortifications, forts, and fortresses of our kingdoms, their situation, figure, and profile, and to know the importance of every one of them, where their strength or weakness lies, whether the lines be drawn to a due length, or the chief angles truly formed.

To make plots, or models of all manner of fortifications, both forts or camps, commanded by us to be erected for our service, and thereof to make the propositions, and to draw estimates of the charge to be considered of by the Master-General of our Ordnance, and Principal Officers thereof, and to be presented to us, for our approbation; he is likewise to represent to the Master-General of our Ordnance, or Principal Officers thereof, the materials requisite and necessary to be therein used and employed, and to see that the same be good and fit for our service, and to cause his assistant or master workman under him to instruct the workmen employed how to use and handle with advantage their shovels, spades, mattocks, wheelbarrows, &c., and how to gain time in working. To see likewise that the Overseer or Clerk of the Cheque for the fortifications look unto the tools, and materials, and take care they be preserved, and that what are broken be forthwith repaired and amended if so it may be, and as soon as the work is done to cause them all to be returned into our magazines, and to take care that they keep an accompt not only of the workmen, but also of the money paid them for their wages.

As often as he shall be commanded by us, or directed by the Master-General of our Ordnance, to visit all the fortifications of our kingdoms, and to make his report in writing, of the condition he finds them in, and exhibit the same to the Master-General of our Ordnance, to be by him presented to us, and to cause the draughts or designs thereof to be left in the Office of our Ordnance, there to remain for the use and information of our said Master-General and Principal Officers of our Ordnance, as occasion shall require.

That when any new designed fortification is by us commanded to be set out, he see the same done, either by himself or the chief engineer next him, and to be present at the beginning or laying of any part of the fortification, and when any undertaker or undertakers have contracted for carrying on the works, either in part or in whole, he see that the same be performed according to the articles, and conditions of their contracts, and when they have finished their undertakings, to make a measurement thereof, and certify the same to our Master-General, and Principal Officers of our Ordnance under his hand.

To endeavour to provide for our service good and able engineers, conductors, and work bases, and not to recommend any to the employ of an engineer but such as are sufficiently qualified to be so, which that it may be known he is to examine what skill the person that sues to be employed hath in the mathematicks, and particularly in fortifications, what works he hath undertaken or managed, in what campaign he hath served, at what sieges he hath been, the manner of trenches, and of the offence and defence, and having gone through this, or such like examination, then to give his report thereof in writing to us, or the Master-General of our Ordnance.

In time of action, or when there is intention of forming or laying a siege against any place, he is to have a draught or ground plot of the place if possible; if not, to take a careful view of its situation as near as he can, and thereof to make a draught, and to see where the attack or attacks are most advantageously to be made, how the circumvalation, and contravalation (if need be) is to be laid out and designed, and to direct, and see the breaking of the ground, planting of batteries, making of platforms, conducting of trenches and mines, and to leave such engineers, and conductors, as will be necessary to see them carried on and executed, to be constantly moving from one attack to another, to see that all possible expedition be made, and so to divide the engineers under him that they may relieve one another, and never to suffer (as far as his authority extends) any single person to be wholy entrusted with a work, or an attack, without he be well assured of his ability and capacity to undertake and discharge such a service.

Instructions for the Duties of the Inferior Engineers.

They are to endeavour the improvement of their knowledge in all things belonging to an engineer, and to render themselves capable in all respects for our service, by attaining to the skill of several particulars mentioned in the first article of the instructions of the duty of our Chief Engineer.

To observe and obey the directions of our Chief Engineer in all things relating to our service, either at home or abroad, as far as our said Chief Engineer shall be by us impowered, or commissioned by the Master-General of our Ordnance to act.

To show their designs to our Chief Engineer, and to take his advice and assistance therein, and to receive his judgment whether (upon examination) they may be thought sufficient or fit to be presented to us or the Master-General of our Ordnance.

Proper Duties of our Master Gunner of England.

To profess and teach his art to our under gunners in the exercise of shooting of great ordnance, mortar-pieces, &c., in such publick place or places as by the Master of our Ordnance shall be allotted, and appointed for that purpose, and therein to exercise them, once a month in winter, and twice every month in summer.

To certify to our Master of our Ordnance, or our Principal Officers, the ability of any gunner, or scolar by him instructed, and not to recommend any but such as he knows to be sufficiently qualified and experienced in the art of gunnery, or have done service abroad, and thereby rendered themselves capable to be entertained by us, to be preferred by the Master-General of our Ordnance to our service, nor to recommend any person that shall continue a clerk to any of our officers, or a servant to any other person to the fees of any of our established gunners.

To keep a register of all the gunners that receive any fees from us, their names, persons, professions, places of abode, and to exhibit the same at the Board of our Office, there to be kept by the Clerk of our Ordnance, that so quarterly musters may be made, and none entered upon the quarter books of our office but such as shall appear to our said Clerk of the Ordnance to be really and actually employed in our service.

To keep likewise an exact list or register of all our great guns, as well brass as iron, belonging to any of our ships, forts, castles, block-houses, or garrisons, or remaining in any of our magazines, expressing their several natures, weights, lengths, and dimensions of metal, at the base, tronions, and muzle, and the same to have in readiness, to be presented to the Master of our Ordnance, or to our Principal Officers at their Board whenever it shall be required.

Lastly the said Master Gunner is hereby strictly commanded and required not to take, demand, or receive any fees, other than the ordinary, antient, and usual fees followiug, viz.:

	£	s.	d.
For the examination, trial, and certificate of any gunner	0	10	0
For taking the oath, and for the copy thereof ...	0	5	0
For admittance, and entry upon the Muster Roll	0	2	6

The Duty and Charge of a FIRE-MASTER.

To exercise, once or twice a week, those that he hath under his command, with a great and small mortar piece, to cast stones, bombs and granades filled with sand, at several distances, and to keep them continually at work in the laboratory, to make unfilled paper rockets, to prepare saltpetre, brimstone, and powder for service, to make all manner of covering of filled fire-balls, with lines or cords, to show them how to fill, fit, and apply a petard for several occasions, and of several sorts.

To be always ready to go upon any expedition in our service, and to give always a true and sufficient note or estimate and time of all such materials as shall be necessary upon any expedition or upon any occasion for fireworks, either of war or pleasure, and in the intervals of time to be always providing with a sufficient quantity of unfilled rockets, water and air baloons, that all things be ready to fill and fit upon occasion.

That he honestly and well do learn and inform such inferior fire-workers or scolars as shall be required by the Master-General, or Principal Officers of the Ordnance to teach and instruct in any part or parts of the whole art of fireworkers for war or pleasure, that he or they so learning may be able to give satisfaction of their proficiencies in the presence of the Right Honourable the Master-General of our Ordnance, Lieutenant-General, or the officers that shall be appointed to assist at his or their examination. That he takes especial care that none of our material or ingrediences be embezled, or converted for other uses, but what is for our service, and what he can show an order for the same.

That none of our materials, or ingrediences of what nature soever, be made use of for any learner or new beginner, without leave from the Master-General, and an order from the Board for the same.

To keep his laboratory in a neat and good order, and to see that every night the laboratory, when there is work, be sprinkled with water and swept, that no dust may remain on the floor, working tables, or shelves.

That he endeavours that none be admitted a fireworker or petardier, but who can give good account of himself in his examination, before His Majesty be put to any charge of a proof.

The Duty and Charge of a FIREWORKER *or* PETARDIER.

That he be obedient to all directions he shall from time to time receive from the Chief Firemaster, or, in his absence, his chief mate.

To be very careful in the laboratory, that the same be kept clean, and to prevent any whosoever to smoak any tobacco in or near the laboratory.

To be ready upon all occasions and expeditions for our service, and to prevent all embezlements of our material and ingrediences for fire-works.

Proper Duties of our PROOF MASTERS.

They are, with the oversight of our Surveyor-General, or other our Principal Officers, to make proof of all our ordnance, ammunition and arms, defensive or offensive, and to cause our mark to be set only upon such as endure the full proof, and to expect wages, and reward from us only, by the merit of their service, and to take nothing from the deliverers of the provisions, arms, or stores, upon pain both of losing their places and receiving such further ignominy and punishment as the inconveniency that may grow by their corruption shall deserve.

That one or both of our said Proof Masters, now and for the time being, be very well skilled and knowing in the manufacture of great ordnance, small guns, powder, &c.

That one or both of them do sign, under the Principal Officers of our Ordnance present at any proof, the certificate or report of the respective proofs by them made, and of the particulars of the arms, stores and provisions marked by them with our mark, before the said certificate or report of any proof be entered or registered in the Office Books, for the better manifestation of their own duties, and discharge of the trust of our Principal Officers concerned in the said proofs; and it is further ordered that no proof of any ordnance, arms, powder, or other ammunition and stores whatsoever be proved by the said Proof Masters without appointment and direction of the Master-General of our Ordnance or Principal Officers of the same, injoining the time, place, and manner of the said proofs.

WAGGON MASTER *to the Office of the Ordnance, his Duty.*

He is to take charge, care, and oversight as well of the waggons and carriages for the necessary service of us, our Royal Consort, and the Officers of our Court, attending us in our several progresses, as for the use of our Office of the Ordnance, and Train of Artillery, and to provide that they be kept in good repair, and fit to be employed upon all occasions in our respective services before mentioned, by representing timely such defects as shall happen to any of them, to the Surveyor of our Ordnance or the rest of our Principal Officers, that care may be taken for their amendment.

When the train of artillery marches, he is to look to all the waggons, carts, and carriages thereto belonging, and of the drivers and carters, to order the loads, and marches of the waggons and carts, and to see that the conductors appointed to attend the said waggons of ammunition, and carriages of stores, do their duty.

He is to see good order both in the marching and lodging of the waggons, carts and carriages, that they clog not up the way, nor hinder one another in marching, and to provide that the carters, wheelrights, carpenters, smiths, and such other artificers as ought to attend them, do suddenly repair and amend any waggon, cart or carriage that shall happen to break or receive any damage in marching.

Lastly, he is carefully to forecast that there be nothing wanting concerning his charge when the train of artillery shall be ordered to march, or if there be any want he is to acquaint the Master-General of our Ordnance, or Lieutenant-General, and the rest of our Principal Officers therewith, and to have order for redress.

Proper Duties of the CLERK OF THE CHEQUE *of the Works, Workmen and Labourers in our Tower of London, upon accompt of the Office of our Ordnance.*

To see the workmen and labourers daily employed in the necessary works of the Office, do give them due attendance, and to keep a Cheque Roll of all such as loiter or absent from their duties, and to cause such as are present to answer to his call, which he is to make four times a day (that is to say) at their coming on to work, and going off in the morning, and at the like time in the afternoon, he is likewise to keep a list of all such as are hired to labour, and to see that they, as well as the ordinary labourers, come to work early, and to set down the time when they set on, and how long they continue, and to admit none to be employed but such as shall be ordered by the Office, or by our Surveyor thought needful to be employed.

With the assistance of one of our Surveyor's sworn clerks, to take a view of, and to keep a daily accompt of all such materials and provisions as by any artificer, whether master smith, master carpenter, cooper, bricklayer, or plaisterer, mason, or others employed in the days work of the office, and to have a regard that no artificer, workman, and labourer be entered for longer time than they do serve.

To take strict notice of every cart that brings in provisions and stores upon our accompt, as likewise of every cart that carries any stores out, and to set down or tally the several turns, and to note upon what particular services the said carts are employed, and from what places they come, and to what places they go.

To render a weekly accompt into the Office in two distinct books, one for our Surveyor, the other for the Clerk of our Ordnance, of all and every services before mentioned, setting down the names of the several persons employed, the time of their being upon duty and service, the several materials and provisions by any of the artificers aforesaid brought in or employed, either in the buildings, reparations, or other days work of the office, that so the book given into the Clerk of the Ordnance may every Saturday be viewed and signed by our Principal Officers (whereof the Surveyor and Clerk of our Ordnance to be always two), as is more fully prescribed in the 12th article of the General Instructions.

Lastly to observe such further directions in any thing relating to his employment as shall be given him by our Principal Officers or our Master Surveyor as occasion shall require.

Duties of the Purveyor *to the Office of our Ordnance.*

To observe and follow the orders and directions that he shall from time to time receive from our Master-General of our Ordnance, or Principal Officers of the same, in the taking up, and providing such ships, vessels, lighters, and boats, as our affairs in our said office shall require, and to bring the owners or masters thereof to the Board of our said Office, there to contract at the reasonablest rates that may be for the transportation of such stores and provisions of war, as shall be ordered to be delivered from our said office, for supply of any of our forts, castles, garrisons, or towns, and plantations abroad beyond the sea.

That when any such ship or vessel is hired by our Master of our Ordnance, or Principal Officers of our Ordnance, for transportation of our stores as aforesaid, he shall take care to see that no delay be made in the shipping the same, and if any such neglect or delay shall happen, he immediately to make complaint of the reason thereof, to our Master of our Ordnance or Principal Officers, that speedy course may be taken for the preventing the charge or demurrage or other charges that may by such default fall upon our said Office.

And when any considerable stores shall be shipped from our said office for Tangier, or any other our principal garrisons or magazines, he shall take care to see the said ship fall down the river, and not omit any the least opportunity of wind and weather, to make the best of their way into the downs, or where else they shall be ordered, and not to leave any such ships untill he hath seen them clear and sailed beyond Gravesend, giving strict charge to the master thereof not to make any stop or stay anywhere till he shall arrive at the place he shall be ordered to.

Duty of the MESSENGER *to the Office of our Ordnance.*

To attend duly and carefully on office days, upon the Master-General of our Ordnance and Principal Officers of the Board of our Office, and to observe and follow all such orders and directions as he shall receive from our said Master-General or the Principal Officers, and carry all such letters as shall be written on post days or other days for our service, and carefully to see the same delivered at the post office or otherwise as they shall be directed, paying the usual postage for the same and bring it to accompt.

To carry out all warrants signed by the Principal Officers of our Ordnance for stores and provisions of war to such artificers as the same shall be directed to, to come to the Board and contract for the same at the best and reasonablest rates for our service, as the occasion of our affairs in that our office shall require, and when the same are contracted for, he shall frequently call upon those artificers to hasten and dispatch the said provisions into our stores, as the emergency of our service shall require.

To give summons to all persons whatsoever, that he shall have directions from time to time from the Master of our Ordnance or the Principal Officers of the same, to attend the Board, and to execute all warrants that he shall receive from time to time signed by our said Master of the Ordnance, or the Principal Officers, for taking any persons into custody for misdemeanours or otherwise, and to keep them in safe custody untill such time as they have given satisfaction, and paid their customary fees, and shall be discharged from that detainment by the order of our said Master of the Ordnance or Principal Officers of the same, and not otherwise.

To inform himself from time to time of the market prices of all such stores and provisions of war, as there shall be occasion for the recruiting and furnishing our stores, and to give in the lowest rates thereof to the Principal Officers of our Ordnance, to the end they may the better contract for the same at the reasonablest rates for our service.

The General Duties of the UNDER MINISTERS *and* ATTENDANTS.

They are to perform their services diligently and faithfully, obeying the authority and directions of the Master-General or the Principal Officers, and following the antient and allowable precedents of the office, and not taking upon them to make warrants, bargains, or bills of payments, or to buy, take up, provide, deliver in, receive, or issue, any provisions or materials, or to take surveys, remains, or accompts, or to give allowance for charges, waste, or expenses, or do anything that is proper for a principal officer otherwise than as their assistants, or servants, and signification, warrant, order, or other direction from the Master or Principal Officers at the Board, according to their direction, and the trust reposed in them, for which the said officers must answer, if prejudice come thereby to our service, so as henceforth no clerk, Master-Gunner, Proof Master, or other inferior attendants upon the office or officers shall be allowed any bills, or payments for emptions, nor suffered to serve into our stores any arms, munition, stores, or other provisions (but by order of the Master, or the Board) whereby any suspicion may grow, that he payeth more than is fit, or more than once.

The Duties of the TRADESMEN *and* ARTIFICERS.

In regard of the benefit and privileges of our service, they are to take special care they do their work substantially, and at reasonable prices, and to serve in no bad stuff, materials, or provisions whatsoever.

That no grant, signification, or warrant for any tradesman or artificer's place shall be esteemed as a claim, or pretence to serve in any stores, but of the best sorts and the best cheap, whereby we may lose the benefit and choice of the market, but that our Principal Officers are at full liberty to buy everything at the best hand for our service, if any of the tradesmen or artificers belonging to the Office of our Ordnance shall not readily serve of the best kinds, and at the lowest rates.

That every merchant, tradesman or artificer serving in goods and provisions to our stores, shall within ten days after the full delivery thereof bring his bill for the said goods delivered to the Storekeeper, and from thence within ten days all such bills to be sent to and allowed (that they ought so to be) by the Surveyor, that so debentures may be passed within one month's space after the delivery of such goods and provisions into the stores by the Clerk of the Ordnance; and if any merchant or artificer shall neglect the passing of his bills according to the rules before mentioned, every such person so offending shall be dismissed from serving provisions any more into our said Office.

Lastly.

Whereas we are given to understand that upon the exposing to sale of old and decayed stores, the clerks and other ministers of the Office are solicitous to buy and contract for the said stores, whereby occasion may be given of censuring the office and officers as if they made the sale of the said stores at under rates, and in favour of the said clerks, which practice (if continued) may be of very ill consequence, and prejuditial to our service, and that the Master of our Ordnance hath declared to us his dislike of the said practice, our will and pleasure therefore is, that for the future no old, decayed, or unserviceable stores shall be sold to any clerk or other under officer belonging to our Office of the Ordnance, or any for them, but that the same be sold to such as shall bid most, and at the best market that may be had, and most advantageous for us. And our further will and pleasure is, that no artificers belonging to the office shall buy any decayed stores or provisions whatsoever, being of the same nature with those he usually serves into the office, for the better avoiding any abuse which may be offered to our service by revending or bringing back into the stores any of our said provisions.

"AMENDED" INSTRUCTIONS.[1]

James II. His majesty's warrant to Lord Dartmouth, master-general:

Whereas, for the better carrying out of our service in the office of our Ordnance, we have thought fit to make several additions and amendments to the "INSTRUCTIONS"[2] of our late dear and entirely beloved Brother, which we hereby confirm, our will and pleasure is, and we hereby authorise and require you, the present and the succeeding Masters of our Ordnance for the time being, to see that the several Articles hereafter mentioned be duly and punctually observed by all Officers and Ministers who may be therein respectively concerned, anything in the said Instructions to the contrary in anywise notwithstanding. And our further will and pleasure is, and we do hereby authorise and require you, the Master-General of our Ordnance, that you demand and receive into your charge, for our service, all warrants, draughts of fortifications, reports, certificates, and all other papers whatsoever relating to the office of our Ordnance from the executors or other the relations of Sir William Compton, Colonel William Legg deceased; as likewise from Sir John Duncombe, Sir Thomas Chichely, and the late Commissioners for executing the place of Master-General of our Ordnance; also from the executors of Colonel David Walter, late Lieutenant, and Sir Bernard de Gomme, late Surveyor of our Ordnance, or any other of the Principal Officers deceased; and that a constant care be hereafter taken by you, the present and all succeeding Masters of our Ordnance, and the Board duly to observe and put in practice what is by us hereby required relating to such papers, &c., as

---

1 It has been deemed desirable to insert here the alterations subsequently made in the foregoing instructions, so as not to interrupt the narrative portion of these notes.

2 See page 52 *ante*.

aforesaid that shall be remaining in the possession or keeping of any Master, Commissioners, or Principal Officers deceased or otherwise removed, which said papers we will be carefully kept in the archives of the Office of our Ordnance for the service of us and our Royal successors, hereby strictly charging and requiring all persons that herein are or shall be concerned to yield entire obedience to these our commands, upon pain of our high displeasure; for all which this shall be to you, him, or them that shall succeed you a full and sufficient warrant.

What is to be done in the Master-General's absence.

Forasmuch as the Master-General of our Ordnance for the time being may be upon occasion by us employed in some expedition for our service, either by sea or land (as lately our present Master-General of our Ordnance was), now to the end our service in his absence may not any ways be prejudiced, retarded, or obstructed, thro' the want of such significations, directions, or orders, as are by us upon any service required to be by him given, we hereby will and require that in such case our Principal Officers, or the major part of them (whereof the Lieutenant-General of our Ordnance for the time being to be always one), by common advice and consent of most voices, and especially by following the literal prescripts of such warrants, as they shall receive either from us, the Lords of our Council or Lord High Admiral, or Commissioners for executing that place for sea affairs, do contract for, and make provision of, receive, and issue any stores or habiliaments of war requisite for our service, and enjoyned by warrants as aforesaid, as likewise that they order lists to be made for the payment of any money thereupon due, and received into the treasury of our said Office, to be signed by three of them at the least (the Lieutenant-General always to be one), and that they do and perform any other service incident to the said office during such time as the Master-General of our Ordnance shall be absent. Provided that due entry be made in the journal of the Office of all warrants received, and of what is or shall be by the said Officers thereupon ordered or directed, and how the same hath been executed and performed, as likewise that an abstract or minute-book of all warrants, orders, and directions entered in the journal of the Office as aforesaid be drawn from thence and signed by the said Principal Officers, or the major part of them, and exhibited to the Master-General of our Ordnance at his return, the better to inform him how matters have been carried on during his absence.

Whereas we are given to understand that the keeping journals and leidger-books by our Surveyor-General, as books of cheque and comptroll upon the accompts of

The Surveyor to keep Journal-book only.

the Treasurer and Storekeeper of our Ordnance, will be matter of charge, and rather impede and hinder than be of any use or advantage to our service, in regard to the necessary cheque and comptroll upon the said accompts are sufficiently provided for in the particular Instructions to the Clerk of our Ordnance, and by the examination of the journal and leidger books, by all or the major part of the Principal Officers, our will and pleasure therefore is, That our Surveyor-General do keep a journal only, and not any such leidger book, as he is by us prescribed and enjoyned do, by way of cheque, and comptroll upon the Treasurer's and Storekeeper's accompts, in the particular Instructions for the common duties of our Principal Officers of our Ordnance.

Storekeeper to make out bills in part of warrants.

Whereas by the second Article of our Storekeeper's proper duties, he is thereby enjoined to receive no stores whatsoever into our magazines, so as to be charged therewith, upon accompt, without warrant in writing signed by our Surveyor and Clerk of our Ordnance, and when such warrants are compleated then to give a bill of receipt of the total of stores thereupon brought in: our will and pleasure is that our said Storekeeper do not receive any new stores into our magazines as to be charged therewith upon accompt without warrant signed by our Surveyor and Clerk of our Ordnance, and grounded upon signification from the Master-General of our Ordnance, or otherwise than as is prescribed in the said second Article of our Storekeeper's proper duties, but if our service shall require we will that our Storekeeper do, by order of the Master-General or the Board, and not otherwise, give bills of receipts for the proportion of stores respectively that shall from time to time be served into our magazines in part of any such warrants as aforesaid untill the same be finally compleated.

Storekeeper: how to be informed of the value and price of stores allowed upon bills.

For the better enabling our principal Storekeeper to keep up his books of accompts, valued according to the true rate of every particular species of stores, as is already enjoined in the proper duties of our said Storekeeper, we will that all stores that were in the magazines of our Tower of London, taken in the last General Remain, 1683, be particularly valued by our Surveyor-General, and approved of by the Board of our Office, to be communicated in writing to our said Storekeeper, with the respective values of all stores that have been or shall be returned from sea or otherwise, and the prices of all other stores that have been brought into our said magazine since the said General Remain, and our said Surveyor is also to take care for the future that the bills for any stores that shall be brought into our said magazine, after allowance thereof by him made, be delivered to our said Storekeeper or one of his sworn Clerks, that the true value may be entered in the books to be by him kept, and the bills immediately redelivered to the Surveyor, or one of his sworn Clerks, that there may be no pretence of delay to the passing the same into debentures according to the course of the Office.

Journal to be kept.

And whereas the particular method and form prescribed our principal Storekeeper for the keeping of his leidger-books in the foregoing Instructions is in some respects more suitable to that of a journal, we will and require that his journal be kept accordingly, and that his leidger be agreeble to the specimen exhibited to the Board, and then concluded upon by the Principal Officers of the same, the 16th of September, 1684, and now in practice.

Monthly audit.

That the Principal Officers do cause to be produced monthly (or oftener if required), at the Board of the Office, all such proportions as have been by them signed, and delivered to the Storekeeper for the issuing any stores and provisions of war, as well for sea as land service, that so it may be seen upon what warrants or orders the said proportions were grounded, how the same have been executed, whether wholly or in part, and if in part, what species of stores were wanting, and not delivered. And that it may the better appear, the Clerk of the Deliveries is required, at the same time, to produce the indents by him taken upon every such delivery, that both the indents and proportions may be compared together by the major part of the Officers at the Board, and by them noted whether the said proportions have been so fully complied with, as they ought, or only in part compleated, which difference (if any shall happen) shall be attested under the hand of the major part of the said Officers, at the time of the examination of the respective proportions and indents at the Board, causing upon every such examination the particular proportions and indents to be distinctly numbered in course, and marked accordingly upon the back of each, which number and marks are to be entered by the Clerk of the Ordnance in the journal of the Office, that so it may appear what proportions and indents have passed examination, and that the Clerk of the Deliveries do first attest upon the back side of every proportion what of the said proportion hath been delivered, at what time, and to whom.

That no piece of ordnance be mounted or dismounted, nor wheels, extrees, &c., be applied for repairing of carriages, &c., nor the property of any stores whatsoever altered without the knowledge of the Board, so that the Storekeeper may be duly charged and discharged, and that no stores whatsoever under the actual charge of the principal Storekeeper be at any time displaced or removed without his particular knowledge.

That an accompt of the expense of nails, oil, marlin, and other petty emptions used about the stores for our service, and the said expense being attested by the Clerk of the Deliveries, or one of his sworn clerks, be brought every month to the Board, that a proportion may be ordered and drawn for the discharge of the Storekeeper.

Periodical sale of stores, &c,

That, as occasion shall require, and as often as the Board shall find necessary for our service, an order may be made and a proportion drawn for clearing the storehouses of unserviceable stores, to be first carefully surveyed, and then disposed of, by way of sale or otherwise, for the best advantage of our service, that so there may be convenient room for the receiving of new stores.

Remains: when to be taken.

That a remain of our stores within the custody of our principal Storekeeper be begun and taken the 10th March next, by the Officers of our Ordnance, or the major part of them, signed under their hands, and entered in two books, the one to be exhibited to the Master of our Ordnance, to be presented to us for our Royal view, the other to remain in the Office of our Ordnance, as a charge for our Storekeeper's yearly accompting according to his Instructions, and that from thence successively every seven years a general remain of all our stores be taken by the Auditors of our Imprests, assisted by our said officers, to the end that an exact state and balance of stores may be made with our respective Storekeepers, and at any time within the space of the seven years, such other remain, or remains to be taken by our Principal Officers, as the Master of our Ordnance shall direct.

And the said auditors are hereby required, upon the yearly examination of the accompts exhibited by the Storekeepers of our Ordnance, armory, small guns, and saltpetre (as well as those of the treasurer of the said office), to give such timely and necessary dispatches to the respective accomptants, upon their regular and just performances, as may tend to their legal and proper discharge for the same.

That the Storekeepers of our out-ports and magazines do keep their leidgers of stores duly and fully posted, cast up and stated, with the rates and value of each particular specie, as the leidger of the principal Storekeeper of our Ordnance is enjoined to be kept; and for the better enabling

The Storekeepers: how to make their accompts.

them so to do, the Surveyor of our Ordnance, and other the Principal Officers whom it may concern, are to take care that they be duly furnished with books, or other informations of the particular rates and values of all manner of stores under their charge, as the same shall from time to time be changed and altered. And they are also hereby required, from time to time, to give the like accompts to the Board of our Ordnance as they are by us enjoyned and prescribed to do, to the Master-General of our Ordnance, in the Instructions for the Storekeepers of the out ports and magazines; also to obey such orders and directions as they shall from time to time receive from the Board of our said Office.

And whereas we are given to understand, that it is in many respects more convenient for our service, that our stores of saltpetre be kept apart, and distinct from those under the charge of our Storekeeper-General, as having never been mixed with the common stores, nor committed to his custody, tho' in the time when the King was served with saltpetre of the English manufacture, and for that the present Storekeeper thereof, Edward Hubbald, gentleman, hath employed for 16 years past, and managed the same with approved judgment and much integrity, having likewise given considerable security for the just and faithful performance of the said service, our will and pleasure therefore is that the same person be continued as storekeeper of our saltpetre at the Tower, Woolwich, and elsewhere, and that you make to him an allowance of sixty pounds per annum to be paid upon the quarter books of our office, causing him diligently to observe and regard the following instructions, viz.—

That he and his successors in the said employment take care of all saltpetre provided for our service, and committed to his or their charge and custody, and to indent and put in such legal security for the safe keeping the same, and for making and giving in a true accompt thereof from time to time, as the Master of our Ordnance, or the Board shall think reasonable.

Saltpetre to be a separate charge.

That he or they receive no saltpetre whatsoever into our magazine without order, or the Master of our Ordnance, or Principal Officers, nor untill the same have been viewed and weighed off, in the presence of the Surveyor and Clerk of the Ordnance or their sworn clerks, and an accompt thereof both as to quantity and quality entered in their respective journals.

That, he or they issue or deliver no saltpetre under his or their charge, without a proportion in writing signified by three or more of our Principal Officers of our Ordnance, nor without indenture thereof taken by the Clerk of the Deliveries and the receipt of the person or persons to whom delivered taken by the said Storekeeper in his leidger book for his discharge, and that upon all such issues or deliveries there be present the Surveyor, Clerk of the Ordnance, and Clerk of the Deliveries, or their sworn clerks in case all the said officers cannot be present.

That he, and they, keep two journals, one of receipts and one of deliveries, and a leidger-book deduced from the said journals, by way of debitor and creditor, that so the Master of our Ordnance and Officers of the same may, from time to time, be readily satisfied what is or ought to be the state of our saltpetre, what hath been received and what thereout issued, and for what services.

Travelling bills.

That no bill be allowed or debenture made for travelling charges or expenses to any person unless he be appointed by the Master-General of our Ordnance, or the Board, and at the same time the day of his going upon our service be by like appointment entered in the journal of our Office, and that the first office day after his return, he do represent to the Master or the Principal Officers at the Board of the Office the service he has performed, and how long he has been so employed upon our service, that the time when the allowance for travelling charges ought to cease, may be at the same time entered into the journal of our said Office.

That the usual travelling charges to any the Principal Officers, engineers, or others sent upon occasion of our service to any of our forts, castles, or garrisons, be not allowed to them residing in any place longer than ten days, but in case of longer stay in any one place, that then his or their allowance shall be considered by the Master-General of our Ordnance, and the Board, and ordered as shall be judged necessary for the services performed and no otherwise.

Tents and Toyles.

Lastly, whereas by our warrant of the fourth of October, 1685, we did authorise and direct that the officer of our tents and toyles should be put under the inspection and regulation of the Office of our Ordnance, and be and remain for the future united thereunto, under the command of you the Master-General of our Ordnance: our will and pleasure is, that the tents, toyles, waggons and other the necessaries to the said office belonging, be committed to the charge and custody of our Principal Storekeeper of our Ordnance, to be at all times kept in such condition as may be fit and ready for our service when commanded. And our further will and pleasure is, that you continue and employ in the Office of our Ordnance, to be borne upon the establishment thereof, no more or other than one yeoman of the toyles, after the rate of £20 per annum, and we likewise authorise and direct that only two labourers fit to attend that duty be borne upon the establishment of our said Office of the Ordnance, without other settled fee or allowance than whensoever our service shall require their travelling abroad, and then to be allowed both to the said yeomen and the said labourers such daily charges as you the Master of our Ordnance or the Board of our said Office shall think needful.

By His Majesty's command,

SUNDERLAND, Pr.

An Establishment of the Annual Payments and Allowances to be made upon the Quarter Books of the Office of our Ordnance.

	£	s.	d.
To the Master-General of our Ordnance for the time being	1,500	0	0

To the Principal Officers of our Ordnance.

Note		£	s.	d.
Note that these salaries are given in full of all wages, allowances, demands, and pretentions, whatsoever upon the Quarter Books of the said Office of our Ordnance.	Lieutenant-General	800	0	0
	Surveyor	400	0	0
	Clerk of our Ordnance	400	0	0
	Storekeeper	400	0	0
	Clerk of the Deliveries	300	0	0
	Treasurer of our Ordnance	500	0	0
	Secretary or Chief Clerk to the Master of our Ordnance for himself and Under Clerks ..	200	0	0

Attendants on the Principal Officers of our Ordnance.

		£	s.	d.
To the Lieutenant	1st Clerk	60	0	0
	Another Clerk	40	0	0
To the Surveyor	1st Clerk	60	0	0
	2nd do.	60	0	0
	3rd do.	40	0	0
To the Clerk of the Ordnance.	1st Clerk	75	0	0
	2nd do.	65	0	0
	3rd do.	50	0	0
	4th do.	50	0	0
To the Storekeeper	1st Clerk	60	0	0
	2nd do.	60	0	0
	3rd do.	60	0	0
To the Clerk of the Deliveries	1st Clerk	60	0	0
	2nd do.	50	0	0
To the Treasurer of our Ordnance	1st Clerk	60	0	0
	2nd do.	60	0	0
	3rd do.	50	0	0

	£	s.	d.
Deputy Keeper of the Armoury	60	0	0
3 Armourers, each at £26	78	0	0
Armourer at Whitehall	20	0	0
Keeper of the Small Guns	80	0	0
Furbusher at the Small Gun Office	30	0	0
Storekeeper at Chatham	120	0	0
Storekeeper at Upnor Castle	80	0	0
Storekeeper at Sheerness	80	0	0
Storekeeper at Tilbury	100	0	0
Storekeeper at Woolwich	40	0	0
Storekeeper at Windsor	50	0	0
Storekeeper at Portsmouth	120	0	0
Furbusher at Portsmouth	40	0	0
Storekeeper at Plymouth	40	0	0
Storekeeper at Hull	40	0	0
Storekeeper at Berwick	30	0	0
Storekeeper at St. James's Park	20	0	0

Under Ministers.

	£	s.	d.
Principal Engineer	300	0	0
Second Engineer	250	0	0
Third Engineer	150	0	0
Two Ordinary Engineers, or young men to be bred up in the art and knowledge of Fortification, &c.	200	0	0
The Master Gunner of England	190	0	0
His three Mates, each at £45 per annum	135	0	0
Sixty Gunners, each at 1/- per diem	1,095	0	0
Fire Master	150	0	0
His Mate	80	0	0
Four Fireworkers or Petardiers at £40 each	160	0	0
Proof Master, two, each at £20	40	0	0
Waggon Master	100	0	0
Clerk of the Cheque	60	0	0
Purveyor	40	0	0
Messenger	60	0	0
44 Labourers, each at £26	1,144	0	0

1683. 19th Aug. His Majesty's Warrant to cause 3 barrels of fine pistol powder, three hundred pound weight of carbine bullets, 3 cwt. weight of pistol bullets, and 3 cwt. weight of match to be delivered to John Leake, Master Gunner, for the use of the 3 troops of Granadiers (proportionably) which are added to the 3 troops of our Horse Guards, for which, this, with the said John Leake's indenture or receipt for the same, shall be your warrant, &c.

3rd Nov. Sir Christopher Musgrave, Lieutenant-General of the Ordnance,

To his loving friends, the Principal Officers of His Majesty's Ordnance.

Whereas it is required by His Majesty that the number of 61 guns be fired at this his Tower of London for the solemnizing the 5th day of November yearly, these are therefore to pray and desire you to issue, or cause to be issued, such a quantity of powder and other stores as is usual upon such solemnities for the 25 culverins and 36 eight-pounders, making the said number of guns on

Monday next, the 5th of this instant November, and this shall be your warrant, &c.

1683. 13th Nov. A similar warrant for the firing of 61 guns on the 15th November, Her Majesty's birthday.

1684. Jan. Quarter Book for January :

Sir Bernard de Gomme, chief engineer.

Major Martin Beckman, second engineer.

Thos. Phillips, third engineer.

To Thomas Culpeper, Esq., and Richard Wharton, Esq., for their encouragement to travel into foreign parts to perfect themselves in the art and knowledge of fortification and other parts of the mathematicks as may render them capable to serve his majesty as engineers.

Capt. Ernest Henry De Reus, chief fire-master.

Isaac Neilson, his mate.

To Magnus Henry De Reus, James English, George Brown, and H. Jacob Woolferman, fireworkers or petardiers.

To Francis Smith and Capt. Charles Lloyd, proof-masters.

To Capt. Justinian Sherburne, waggon-master.

60 gunners paid quarterly at 1s. each per diem.

1684. 2nd April. His Majesty's warrant appointing Lord Dartmouth Master-General of the Ordnance.

An Account of Harness requisite for the old Battering Train of Artillery in the Tower.

Nature of Ordnance.		Length feet.	Weight.	No. of sets.	Trace Harness.	Thill do.	Jacked halters.	Bitt halters.	Hemp halters.	Number of horses.
Cannon of 8	of	11½	6777	1	28	1	2	27	29	29
do. 7	„	10½	5790	1	20	1	2	19	21	21
Demi do.	„	11½	71 3 15	1	24	1	2	23	25	25
„ do.	„	do.	58 3 28	1	18	1	2	17	19	19
„ do.	„	do.	56 0 0	1	18	1	2	17	19	19
„ do.	„	11	67 3 16	1	22	1	2	21	23	23
24-pounder	„	14	76 0 22	1	30	1	2	29	31	31
do.	„	10	52 3 2	1	18	1	2	17	19	19
do.	„	do.	4746	1	16	1	2	15	17	17
			53 1 2	1	18	1	2	17	19	19
Culverings	„	11½	55 0 10	1	18	1	2	17	19	19
			50 0 0	1	16	1	2	15	17	17
			48 3 15	1	16	1	2	15	17	17
do.	„	10½	45 0 0	1	16	1	2	15	17	17
			48 3 15	1	16	1	2	15	17	17
			46 3 0	1	16	1	2	15	17	17
do.	„	10	40 3 10	1	14	1	2	13	15	15
			39 3 2	1	14	1	2	13	15	15
do.	„	6½	17 0 12	1	6	1	1	6	7	7
Total		19	Guns.	19	344	19	37	326	363	363

REORGANISATION OF THE GARRISONS AND MAGAZINES IN IRELAND.

1684. 24 Aug. By the King's most Excellent Majesty and the Lords of His M.'s most Honble. Privy Council.

The Right Honble. the Lord Dartmouth, Master-General of His Majesty's Ordnance, having made a proposal to his Majesty in Council for several rules and orders to be established in Ireland for the preservation of his M.'s ordnance, ammunition and habiliaments of war in that kingdom under a cheque, that his Majesty from time to time may know the state and condition of the stores and magazines there, and his Lordship having also represented to his Majesty, That whereas at present his Majesty's stores of war in that kingdom are disposed unproportionably in thirty-four garrisons, as appears by the last remain, and that many of them are not only an unnecessary charge but unsafe for his Majesty's magazines to be continued in, and therefore humbly proposing that those garrisons may be reduced to as few as may be with conveniency and safety, and that such as shall be adjudged fit to be continued may be well garrisoned and put into better condition for his Majesty's service, and the defence of that his Majesty's kingdom against foreign invasion or other hostile attempts whatsover, and that all the provinces may be usefully provided for, All which His Majesty in Council having taken into serious consideration was pleased to approve of the said rules and proposals, and to order, And it is hereby ordered by his Majesty in Council that the same be observed and put in execution by the Master and Officers of his Majesty's Ordnance in Ireland therein concerned, according to the directions from the Lord Dartmouth, Master Genl. of His Majesty's Ordnance and Armorys. And it was likewise ordered that copies of the said rules and proposals, which are hereunto annexed, be transmitted to his Grace the Duke of Ormond, his Majesty's Lord Lieut. of Ireland, to the end his Grace may take care that the same be observed and put in execution accordingly as far as his Grace is to give his orders and directions therein; and it was further ordered by His Majesty in Council that the Rt. Honbl. Lord Dartmouth do appoint some fitting Officer of the Ordnance here to go into Ireland to take a remain of the ordnance, ammunition and other habiliaments of war there, according to the first rule proposed by his Lordship; And whereas there was also an estimate presented to his Majesty by the said Lord Dartmouth of the charge of all manner of necessaries to be provided for a train of artillery of twenty pieces of ordnance and two mortar pieces in that kingdom (a copy whereof is also hereunto annexed, amounting to the sum of one thousand two hundred sixty-seven pounds ten shillings; His Majesty in Council upon taking the same into consideration was pleased to order that only one moiety for his Majesty's present service of the said proposed train with necessaries proportionable according to the said estimate should be sent into Ireland; And that the Rt. Hble. the said Lord Dartmouth, Master General of his Majesty's Ordnance, do forthwith take care that the same be done; And to enable his Lordship thereunto, The Rt. Hbl. the Lords Commissioners of His Majesty's Treasury are to give all necessary directions for supplying the Treasurer of his Majesty's Ordnance with the sum of six hundred and thirty-three pounds fifteen shillings out of his Majesty's revenue of Ireland for the said service, being one moiety of the said estimate; And lastly it was ordered by his Majesty in Council that the said Master General of his Majesty's Ordnance do forthwith take care that one thousand barrels of gunpowder which his Lordship has his Majesty's commands to send into Ireland, and for the payment whereof he hath received bills of exchange from his Grace the Duke of Ormond, be dispatched away with all possible speed.[1]

PHILLIP LLOYD.

RULES AND ORDERS.

It being for your Majesty's profit as well as security to have such rules and orders established in Ireland, as may best preserve the ordnance ammunition, and habiliaments of war, under a cheque, that your Majesty from time to time may know the state and condition of your magazines, and stores in that kingdom,—

It is humbly proposed

1st. That some officers of his Majesty's Ordnance in England may be commanded into Ireland to take a remain of all the ordnance, ammunition, and other habiliaments of war that are in any town, garrison, castle, or fort in that kingdom.

2nd. That the Master of the Ordnance in Ireland may be directed to appoint persons to attend taking of the remain, that they as his deputies seeing the account taken, the Master of the Ordnance may be charged therewith.

3rd. The remain being so taken, a copy to be delivered to the Duke of Ormond, your Majesty's Lieutenant, and another copy brought over and lodged in your Majesty's Office of the Ordnance in the Tower, to remain for a cheque to future account.

4th. That the Master of the Ordnance in Ireland do annually transmit a general remain into England to shew all the expense, decays or wants of the year preceding, to the end your Majesty may be truly informed of the condition of the magazines.

5th. That the most proper places for the making of magazines being resolved upon, they be reduced to as few as may be with safety for the lessening of the charge, and other considerable reasons.

---

[1] It will be seen that on the 22nd March, 1687, this amount was increased to £2312 1s., as per detail given at page 107.

6th. That all ordnance and arms where there is no established garrison be removed to the magazines.

Whereas at present your Majesty's stores of war are dispersed unproportionally into thirty-four garrisons, as may appear by the last remain many of which are not only an unnecessary charge, but unsafe, for your Majesty's magazine to be continued longer in, it is humbly proposed that those garrisons may be reduced to as few as may be with conveniency and safety, and that such as shall be adjudged fit to be continued be well garrisoned and put in a better condition for your Majesty's service, and the defence of your Majesty's kingdom of Ireland either against foreign invasions or other hostile attempts whatsoever, and that each province may be usefully provided for, it is humbly offered That in

LEINSTER

Dublin may be continued the chief magazine in respect it is the metropolis of that kingdom and the residency of your Majesty's Lord Lieutenant, the Exchequer and Courts of Justice, But if the Castle of Dublin be not secured enough to be preserved from any surprise or insult, It is humbly offered to your Majesty's consideration, whether the citadel sometime since designed by St. Bernard de Gomme be needful to be built, or if that (by reason of the greatness of the charge) is not found convenient to be undertaken, then that some other fort of less expense may be erected for the safely lodging your Majesty's stores of war near this city

CONNAUGHT.

Athlone is represented as the most proper place for the lodging the train of artillery, and placing a good magazine of arms, ammunition and all utensils of war, not only in respect of its being capable of being made strong, and that your Majesty hath a good castle already for the lodging your stores in, that commands the most considerable pass, and will be as serviceable a garrison for Leinster as this province; But that it is near the centre of the whole kingdom, so that the train and forces to be lodged here, will be most conveniently placed and readiest to march to the succor of any part of Ireland, whenever it shall be needfull.

Galloway is represented to be well situated with a strong old wall about it; there are two small citadels built by the late usurpers which require to be repaired. It is a town of trade with an indifferent harbour, and a good road without it; it is the chief port and key of this province, and conveneint to receive a good garrison, and to lodge your Majesty's stores in.

MUNSTER.

DUNCANNON is a large blockhouse situated upon the river of Waterford, and commands the channel; it is convenient for lodging your Majesty's stores and continuing a small garrison, but WATERFORD is proposed as more convenient for quartering a more considerable number of your Majesty's forces in this province.

LIMERICK is represented as the strongest place in all Ireland, being encompassed by the Shannon and fortified by the late usurpers, is capable of receiving a considerable garrison, and hath a castle convenient for lodging your Majesty's stores in.

CHARLES FORT lately built at Rincorian, was designed by the late Earl of Orrery to be built in the form of a regular pentagon with a crown work, and good batteries down to the waterside; which crown work and batteries are said to be finished; but the citadel being not yet begun, this new work is enclosed with small bastions to the landward, for the present only enough to secure it from surprise or sudden insult: and those batteries already done do in great measure secure the entrance into the harbour of Kinsale, which was the chief end of building this fort, and it will require good consideration before your Majesty be put to the charge which was at first proposed, by reason of the situation as it is represented, being encompassed with several hills which will command it, and the extent so small, if it were finished according to the design drawn, that it would not be long able to resist the force of mortar pieces, great batteries and a regular army, so that what is done already being sufficient to keep it from any sudden attempt, and the river being as much commanded as is intended in this place, it does not seem reasonable to put your Majesty to any further charge in this place at present, but to keep as good a garrison, and lodge as much stores as the place is capable to receive.

It is necessary that another battery be made at the end of the reach near KINSALE for the better security of the town and harbour, which is one of the best in this kingdom for the safety of your Majesty's navy, and therefore ought to be well defended.

There is a fort and blockhouse at CASTLE PARK, which is a peninsula, and of stronger situation, and of greater extent for building a place of strength for lodging your Majesty's stores and arms, for the supply of this part of the kingdom; but this only hinted at, according to the information given me; therefore it is humbly desired that the Master and some of the officers of the Ordnance in Ireland with some able engineer may make a survey, and give their judgments upon the place.

ULSTER.

CHARLEMONT was represented upon a survey taken in 1662 to be no convenient place for making a magazine, but rather that Ardmagh should be fortified, and garrisoned for your Majesty's service, as may appear more at large with the reasons then given; but notwithstanding that said report this house hath been since bought of the Lord Calfield for your Majesty's use, refortified and garrisoned so that there is no other place represented as fit to lodge the stores in, in those parts, and must therefore be continued a magazine, tho' it be capable of receiving but a small garrison; but ARDMAGH lying near, it will be a good quarter for some of the forces in this province, and may be convenient to be fortified hereafter, if it shall be adjudged fit for your Majesty's service; but the TOWER OF CHARLEMONT being represented lately as more convenient to be fortified, and at less charges than Ardmagh; and yet the fort already there may serve as a citadel to the said town if it be fortified, it is humbly proposed that both places may be surveyed by the Master and officers of the Ordnance in Ireland, and with the assistance of some able engineer, and that they may report their opinions under their hands, before your Majesty be put to so great an expense.

LONDONDERRY as appears by the report made in 1662 is a city strongly seated by nature, the great river Lough Foyle encompassing the same on three sides, making here a goodly, capacious, and safe haven: that ground to the landward is marsh, and with small cost may be cut to bring the water quite round the town; it is walled with a good stone wall, rampered and well flanked, in good repair, the defects nothing considerable, save the want of platforms upon the bulwarks. The town is built upon a steep rising ground from the haven; on the highest part thereof (the whole length of the town from the key) is their church, and churchyard, joyning to the city walls, of which the late usurpers made a kind of citadel, including the church and part of the churchyard, making the steeple the magazine; here are some twenty and odd iron guns serviceable in themselves, but all dismounted.

The greater part of the inhabitants of Derry are Scots, as is all the country hereabouts, indeed all Ulster over.

It was moved the magistrates for repairing the few defects, especially the antient storehouse (being a place very convenient for the magazine), making platforms for the cannon, and mounting the same upon carriages, the guns having upon them the marks and arms of the several Companies of London, and delivered an estimate of the charge to the Major, the repairing the storehouse, and the demolishing some poor cottages upon the walls, but the rest he must recommend to his masters of London.

This town is considerable both for strength and situation; it lies upon a noble haven, without bar or let, twenty miles from the sea, into which ships of great burthen may enter without danger, the same flowing and being navigable many miles further up into the land; but the people not industrious, nor well accommodated to make use of the conveniencies the place affords them. A regiment of foot and two troops of horse will be a good garrison in this town, which well ordered will have an influence upon the northern parts of Ulster, preserve the peace, and oblige the people to obedience, and those troops will be of far greater use in this place, than as now they are dispersed in private villages, to preserve private interests.

CULMORE CASTLE lies three miles down the river from this place, a triangular work, well designed in those times to command the channel, the river rising in the form of an half moon about it, and a bog on the land side; in the last troubles there was a large work begun to take in the ground lying between the bog and castle, which may contain three or four acres of land, but this work was never finished, and it was advised to demolish what was raised; here is neither store nor cannon mounted. There is an establishment of £1 9*s*. 4*d*. per day for a captain, a gunner and ten wardours, payed by the City of London; this place if thought fit to be continued, a garrison may be kept by commanded men from Londonderry.

Carrick Fergus, a town seated in a plentiful soil on the eastern shore, within sight of Scotland; it hath a noble road, wherein ships of greatest burthen may ride safely in all weathers without peril; the town is of no great extent, is walled round with stone and well flanked; but the walls and ramparts in some places not well finished; Here is a large castle, three wards or retreats one within the other, and in the innermost a fair large square tower, there hath been good houses and fair cellarages, and rooms for magazines and stores of all sorts; but these have been neglected and suffered to fall into ruin—a great pity and a great prejudice to the Crown's service, this being the best key of Ulster, strong in itself, and apt to receive all things for your Majesty's service, in that province, out of any other part of your Majesty's dominions; here are fifty-seven guns of all sorts, but most unmounted, the greatest number of them are brass; some course ought to be taken for repair of this castle (which else in short time will be past help) and to that effect a carpenter and mason did view the defects, and an estimate of the charge was drawn, and delivered to His Grace the Duke of Ormond.

These are the places most considerable to be garrisoned according to the best information that I have hitherto received; but it is humbly desired that His Grace the Lord Duke of Ormond, Lieutenant, and your Majesty's Council may judge if any other forts are absolutely necessary.

(Signed) Dartmouth.

When artillery was required to accompany the troops in any expedition, a demand was made by virtue of the king's special warrant in each case, "on the principal officers of the ordnance for the number of guns, and the requisite stores, ammunition, and equipages for the same;" and these were delivered over to the officer named, who was responsible for their use. Thus, when King James II. was collecting his army to oppose the rebellion excited by the Duke of Monmouth (16th June, 1685), he issued a warrant to Lord Dartmouth, the Master-General of the Ordnance, directing him to order the storekeeper at Portsmouth to get ready a train of artillery, consisting of eight pieces of ordnance, viz., "four 3-prs. iron, nealed and turned, and four brass falcons,[1] with all such stores, ammunition, and equipage,

[1] "The *falcon* was a $2\frac{1}{2}$-pr., 7 feet long, and weighed 750 lbs. The charge was $2\frac{1}{4}$ lbs. of

as he may judge requisite to attend the same for an immediate march." This warrant was issued on the 16th June, 1685, three days after the Court and Parliament in London received the intelligence of the landing of the Duke of Monmouth, at Lynn. A train of artillery was consequently prepared and despatched to join "*the present expedition under the Earl of Feversham, in opposing the treacherous practices and attempts of those rebels which are now openly in arms against us.*"

1685. The following officers and persons were appointed to this train, with their wages per diem:—

		£	s.	d.
Comptroller..........Henry Sheares, Esq..........		0	15	0
His clerk		0	3	0
Commissary of ammunition for the train and army....		0	10	0
Gentlemen of the ordnance		0	5	0
Clerk to the commissary		0	3	0
Quarter-master		0	5	0
2 conductors	2s. 6d. each	0	5	0
Fire-master		0	5	0
Master-gunner		0	5	0
8 gunners	2s. each	0	16	0
8 mattrosses	1s. 6d. each	0	12	0
		£4	4	0

This was one gunner and one mattross to each piece of ordnance, and these received double pay on service. The guns were thus given over to the care of the regiments with a small proportion of gunners, who had been exercised as above to "charge and shoot them off;" for on a subsequent occasion, when two 3-prs. were distributed to each of seven regiments at Hounslow (8th August, 1686), it is directed that the Grenadiers belonging to those regiments are to convey the field pieces to the appointed place. Of the officers, the firemaster appears to have been the chief artillery officer (if we may use the expression), for his duties are thus detailed in the "Book of Instructions," page 77 *ante*, "To exercise once or twice a week those that he hath under his command, with a great and small mortar-piece, to cast stones, bombs, and grenades, filled with sand, at several distances, and to keep them continually at work in the laboratory, &c. . . . . . . to shew them how to fill, fit, and apply a petard[1] for several occasions,

corn powder. This gun was drawn by three horses, or two yoke of oxen, and had a range from 260 to 1200 paces."—*Animadversions of Warre.* By Robert Ward, Gentleman and Commander, London, 1639.

1 "*Petards* are only used to blow open ports, and portcullis; they are divers in their magnitudes, according to the strength of the opposition they are to mine. Their proportions are,—some to hold one pound of powder and lesse; others to hold fifty or sixty pounds or more. They are fashioned like a morter; or saint's-bell, somewhat tapered. A strong hooke is to be scrued into the substance you intend to mine, and upon this hook hangs the wringle of the petard, and likewise to be shored up with the strong forked rest to stay the reverse of it."—*Animadversions of Warre.*

and of several sorts." His duties in the field were evidently to superintend the troops in the use of the artillery of the train.

On the 21st June, 1685, a second warrant was issued to the Master-General, directing him to have in readiness a train of artillery, "*necessary at this juncture of time, when divers traitors and rebels are in open hostilities against us,*" consisting of sixteen pieces of brass ordnance, with the requisite stores, ammunition, and equipage from the stores in his charge, with all expedition, to march towards Chippenham in Wiltshire, to join the Earl of Feversham, whose orders the officers of the said train were to obey and observe.

The following brass pieces composed this train: viz.

2 Twelve-prs.,	4 Sakers,[2]
4 Demi-culverins,[1]	2 Minions,
4 Six-prs.	Total, 16,

with forty round, and fifteen case shot for each gun.

By the following list it will be seen that two gunners and two mattrosses per gun, was considered a fair proportion:—

List of Officers, &c., appointed to this Train.

Name	Office	Pay per diem. £	s.	d.
Captain Francis Povey ...	Comptroller....................	0	15	0
Thos. Hastler	His clerk	0	3	0
Abell Barton	Commissary of ammunition	0	10	0
James Hadder .... / Jenkin Owen..............	His clerks ....each 3s.	0	6	0
John Arnold	Pay-master	0	8	0
Edward Silvester	His clerk	0	3	0
Robt. Gargrave	Waggon-master	0	10	0
Wm. Barker .... / Joseph Leidgingirm	Assistants .... each 4s.	0	8	0
Thos. Hardiman	Commissary of the horses ...	0	8	0
——— / ———	Assistants ....each 4s.	0	8	0
Richd. Anthony........... / Richd. Elliot .... / Charles Mossom...........	Gentlemen of the ordnance each 5s..................	0	15	0

1 "The regulation *demi-culverin* was a brass gun of 11 feet, weighing 3000 lbs., and projected a shot of 11¾ lbs. with a charge of 12½ lbs. of corn powder. It required 7 horses (or 4 yoke of oxen) to draw it. It had a point-blank range of 380 paces (of 5 feet), and carried at '*utmost random*' or extreme range 1800 paces. (The gun referred to in the text, was probably the *lesser demi-culverin,* which was also an authorized peece. This peece threw a ball of 9 lbs. with 7½ lbs. of corn powder; was 10 feet long, and weighed 2,300 lbs. It was drawn by 6 horses or 3 yoke of oxen)."—*Animadversions of Warre.*

2 "The ordinary *saker* was of the following dimensions:—viz., length 9½ feet; weight 1900 lbs.; weight of ball 5¼ lbs.; charge 5 lbs. of corn powder; range from 300 to 1500 paces."—*Ibid.*

		Pay per diem. £	s.	d.
James Hubbard	Purveyor	0	5	0
Charles Love	Provost marshall	0	6	0
John Caine, William Thomas	Asistants each 2s.	0	4	0
John Fawcett	Master-gunner	0	5	0
Thomas Doage, Henry Hampton	Mates each 3s.	0	6	0
32 gunners	each 2s.	3	4	0
32 mattrosses	each 1s. 6d.	2	8	0
6 conductors	each 2s. 6d.	0	15	0
Delaval Denton	Surgeon	0	4	0
His mate		0	2	6
Tent-maker		0	4	0
Assistant		0	1	6
Rowland Lewis	Master-smith	0	4	0
2 assistants	each 2s.	0	4	0
1 farrier		0	2	0
John Bray	Master-carpenter	0	3	0
4 carpenters	each 2s.	0	8	0
John Archer	Master-wheeler	0	3	0
2 wheelers	each 2s.	0	4	0
John Mitchell	Master collar-maker	0	3	0
1 collar-maker		0	2	0
Phillip Thomas	Master-cooper	0	3	6
1 gunsmith		0	1	6
Edward Hubbard, jun.	Captain of the pioneers	0	4	0
Edward Pateman	Sergeant	0	2	0
Thomas Pateman	Corporal	0	1	6
1 drummer		0	1	0
20 pioneers	each 1s.	1	0	0
118 drivers	each 1s.	5	18	0

Among the stores were the following:—

Falling hooks (pairs)	20
Hurters	40
Ice cart	1
Sarsnet banners for wagons	40
Turnpikes[1]	8
Stoccadoes	200
Men's harness, 67 and 52 to a set of, each	6

The pay per diem of the train was £22 2*s.* 6*d.*

The men attached to the train were armed as follows, viz.,

1 "The use of the *Turne-pike* and how it is framed, and of the excellent defence it makes, both against Horse and Foote upon all straights and passages."—*Animadversions of Warre.*

2 gunners' mates } each a field stave,[1]
32 gunners }
32 mattrosses;[2] each a half-pike, and a hanger with belt,
Serjeant of pioneers; a partizan, and a hanger with belt,
Pioneer corporal; a halbert, and hanger with belt,
20 pioneers; each a hanger and belt, and between them { 8 pickaxes, 6 shovels, 6 spades,
Drummer; a drum, and hanger with belt,
Hangers with belts for the conductors, wheelers, carpenters, coopers, smiths, and collar-makers.

The following was the order of march of the foregoing train of artillery:—

"1st Division: Before the ordnance marched the carts, with spades, pickaxes, and hatchets.

Carts with harness.
Carts with gyns, hurdles, &c.

Before this division marched the captain of pioneers,[3] and his men.

After this as part of the 1st Division, followed the ordnance, the greatest pieces first.

Before the ordnance marched the general, lieut.-general, comptroller, and commissary of artillery.

---

[1] *Field-stave*, "linstock." Shakspeare, in the Prologue to Act iii. of Henry V., has the following:—

"Work, work your thoughts, and therein see a siege:
Behold the ordnance on their carriages,
With fatal mouths gaping on girded Harfleur.
Suppose, the ambassador from the French comes back;
Tells Harry that the King doth offer him
Catharine his daughter; and with her, to dowry,
Some petty and unprofitable dukedoms.
The offer likes not: *and the nimble gunner*
With linstock now the devilish cannon touches,
(Alarum: and chambers go off)
And down goes all before them."

[2] "*Matrosses;* soldiers in the train of Artillery, next below the gunners; their duty is to assist the gunners in traversing, spunging, loading, and firing of guns, &c. They carry fire-locks, and march along with the store wagons, both as a guard, and to help in case a wagon should break down."—*Chalmer's Cyclopædia.*

"Likewise every peece hath his gunner, with his co-adjutor or mate, and a man to serve them both, and helpe them charge, discharge, mount, wadde, cleanse, scoure, and coole the peeces being over-heated."—*Animadversions of Warre.*

[3] "The Master of the Ordnance must have under his command a band of pioneers, the captain over them to be very discreet and carefull; he must likewise see these pioneers be sufficiently guarded whilst they are at worke, lest the enemy surprize them."—*Animadversions of Warre.*

The pioneers attached to the train were clothed in red jackets and red caps, and the artificers in red clothes, laced.

By the side of the train, the master-gunner, and the gentlemen of the ordnance, every one to the division of guns allotted to them; and to every gun the proper gunner, and matross alloted to them.

After the ordnance followed:—

Carts, with budge barrels, cartouches, &c.
Tumbrills,[1] with iron and case shot.
A cart with gyns and matches (*sic*).
Carts with spare camp equipage.
The generals, lieutenant-generals, and other principal officers, wagons.
Wagons of match.
Wagons of powder.
Wagons for tents, as well ammunition tents as others, with the tent-master and his servants.

Then carts of spare material.

Wagons for the smiths, marching forges, tools, &c., and those of carpenters, wheelers and other artificers.

Then followed the 2nd Division, which consisted of the ammunition for the foot and horse, viz.,

Wagons of powder.
,, match.
,, musket, carbine, and pistol shot.
,, spare arms, and all other necessaries.
Bridge boats.

The 3rd Division consisted of the wagons and carriages that carry the standing magazine, and the utensils and provisions that are for the reserve of the train and army above the ordinary daily use."

"Upon all other wagons and carriages, proper conductors are to attend, who are responsible to render an account of all under their respective charges, and how the same are employed and expended."

Of the officers of the train, the wagon-master appears to have held the most responsible office as regards the effective state of the train. His duties are detailed in the "Book of Instructions," page 79 *ante*.

1685. 5 July. The army were encamped at Sedgemoor, where they completely defeated and dispersed the rebels, with the loss of 1500 men.[2]

---

1 From the Spanish "*tambarilla*," a chest or trunk with an arched cover.

2 Monmouth, whose advance was disputed by the militia, at Bridport and Axminster, though he was welcomed at Taunton and Bridgewater, ultimately decided to besiege Bristol; but the Duke of Beaufort was resolved to defend the place to the last, and was enabled to carry out his resolve by the speedy approach of the Royal forces, when the siege was raised. Thence the rebels proceeded into Wiltshire, and on their arrival at Philips Norton, were surprised to find Feversham close at their heels. Despite a temporary reverse, which happened to his advanced guard, Feversham held his ground at Bradford, where he awaited the arrival of the train of artillery. Monmouth fell back upon Bridgewater, where he was overtaken by the Royal forces, under Feversham, with Churchill (afterwards the renowned Marlborough), in a subordinate command. Then followed the battle of Sedgemoor, of which suffice it here to say,

1685. 21 Nov. WHEREAS BY AN INDENTURE heretofore made between King Henry the Seventh and the town of Dartmouth, the said town was obliged to finish, and garnish with guns, artillery and other ordnance defensive and sufficient, a certain tower and bulwarks therein building for the safeguard of the said town and parts adjoining, and from thenceforth for ever to find a chain sufficient in length and strength to stretch and be laid overthwart the mouth of the haven of Dartmouth from one tower to another, and all times thereafter to fortify and repair and keep the said towers and bulwarks garnished with guns, artillery and other ordnance sufficient for the defence of the said chain and port, town, and parts adjoining. In consideration whereof there was by the said indenture granted to the said town for ever the annuity of £40 per annum, payable out of the Customs; and whereas his majesty hath been pleased this day to sign a warrant for a privy seal to confirm the said grant of £40 per annum to the said town of Dartmouth, my Lord Treasurer therefore desires your lordship to give orders to the officer of the ordnance to take care that the said town do duly perform their covenants contained in the said indenture.[1]

To the Master-General of the Ordnance. (Signed) HENRY GUYE.

---

that after the defeat of the rebel cavalry, who commenced the action, and whose retreat caused inextricable confusion in the ranks of the infantry of the insurgents, the Duke of Monmouth, who had hitherto behaved with great bravery, mounted his horse and fled;—not so his infantry, who in spite of the repeated attacks of the Royal cavalry, made a stand until their ammunition failed them. At this period of the action, the King's artillery came up. It had been posted at about half-a-mile distance, but had hitherto taken no part in the action. "So defective were then the appointments of an English army, that there would have been much difficulty in dragging the great guns to the place where the battle was raging, had not the Bishop of Winchester offered his coach horses and traces for the purpose." (*Macaulay's History of England.*) Even when the guns had arrived, there was such a want of gunners, that a serjeant of Dumbarton's regiment (the 1st Royals) named Weems, was forced to take on himself the management of several pieces, and for which he received a gratuity of £40, "for good service in the action at Sedgemoor, in firing the great guns against the rebels." The cannon, though ill-served, brought the engagement to a close, and the rout was complete; the loss of the rebels being about 1000 men, or three times that of the Royal forces. "So ended the last fight deserving the name of battle which has been fought on English ground." This defeat was speedily followed by the capture and execution of Monmouth. Macaulay mentions that the best account of Monmouth's rebellion is to be found in the MS. journal, kept by Mr. Edward Dumner, of the train of artillery, which book is in the Pepysian Library, at Oxford. If a copy of this MS. could be procured for the Regimental Library, it would be an interesting record of the management of artillery in the reign of James II.

1 This seems to refer to a system of subsidy, which finds an illustration in the case of Londonderry, in Ireland; for when a *remain* was taken of the ordnance in that kingdom, as before stated, it was noticed with respect to this town, that most of the ordnance was marked with the arms of the various companies of the City of London. "On the bastions were planted culverins and sakers, presented by the wealthy guilds of London to the Colony (of Londonderry). On some of these ancient guns, which have done memorable service to a great cause, the devices of the Fishmongers' Company, of the Vintners' Company, and of the Merchant Tailors' Company, are still discernible." (Macaulay, Vol. III., p. 142.) "Here and there,

1685. 23 Dec. Unserviceable brass ordnance at Woolwich directed by his majesty to be recast into mortar-pieces or other ordnance, as the master-general may direct for the better completing our several trains of artillery, "for which this shall be your warrant."

1686. Jan. Sir Martin Beckman appointed chief engineer.

June. Jacob Richards appointed 3rd engineer.

12 May. His majesty's warrant directing a train of artillery consisting of 26 pieces of ordnance of 3-pound bullet, with stores, equipage, officers, gunners and other ministers, with all expedition to march and join the army at the rendezvous at Hounslow Heath, there to encamp; and further directing the pressing of teams of horses and drivers for the same, as Thomas Hardyman, commissary of draught horses, may require. [This was a regular train of artillery, subsequently exchanged for 14 regimental guns, as detailed on the 8th August.]

8 Aug. Directions for the disposal of the 14 3-pounder field pieces to the several regiments of foot, by order of his majesty:—

Scotch Guards Royal Regiment Prince George's Regiment Holland Regiment Earl of Bath's Regiment Colonel Kirk's Regiment Marquis of Worcester's Regiment	Tuesday, Hide Park.

MEM.—The granadiers belonging to those regiments are to convoy the field pieces to the appointed places, and the 28 field pieces in camp at Hounslow to be returned to the Tower.

21 Aug. His majesty's warrant: WHEREAS by the great increase of our trains of artillery lately disposed of in several parts of our kingdom for the benefit and advantage of our service, as likewise by the considerable numbers of our ordnance lately mounted for the better defence of divers of our garrisons, and particularly that of our Tower of London, it will be requisite that fitting persons able to undergo the charge of gentlemen of the ordnance should be assigned to take the management, care and oversight of the guns, as well of the respective trains as of our several garrisons: our will and pleasure is, and we do hereby authorize and require you to make choice of, nominate and

among the shrubs and flowers, may be seen the old culverins which scattered bricks, cased with lead, among the Irish ranks. One antique gun, the gift of the Fishmongers of London, was distinguished during the 105 memorable days, by the loudness of its report, and still bears the name of 'Roaring Meg.'" (*Ibid.* p. 240.) Again, at Culmore Castle, where cannon were not mounted, was an establishment of £1 9*s.* 4*d.* per diem for a captain, a gunner, and ten warders, paid by the City of London.

appoint such and so many persons as you shall judge to be able and sufficient to discharge the duties of gentlemen of the ordnance, requisite and necessary to be employed in our respective trains and garrisons aforesaid, without which the ready effects of our artillery in either respect may perhaps be wanting where occasion shall be offered; and we do further declare our royal will and pleasure to be, that for so many as you shall make choice of, and represent to us as necessary to be employed in our service as aforesaid, you cause to be made to them quarterly by bill and debenture out of the treasury of the office of our ordnance, such wages, sallaries and allowances as have heretofore been made to them in like cases, &c.

Captain Smith, Captain Lewin and Mr. Mossome to be gentlemen of the ordnance, as signified by the master-general.

Bombadiers and Petardiers instituted.

1686. WHEREAS we have thought fit of late to order the casting of diverse brass mortar pieces of several sizes, and caused the providing of proportionable numbers of bombs, carcasses and grenado shells [1] to them, and intend more of both for the future, as being very well satisfied of the exceeding usefulness and advantageous execution of that sort of artillery now in practice, according to the modern art of war (more than formerly), and for as much as without the assistance of able persons [2] well qualified and experimented in the practice of the said artillery, little or no use is like to be made of them when our service shall call for the same: our will and pleasure therefore is, and we do hereby authorise and require you to enter upon the quarter books of the Office of our Ordnance as an addition to the officers, ministers and attendants within the said office by the former instructions already established—

One chief bombardier	at 3s.	per diem.
12 bombardiers	at 2s.	do.
One chief petardier ...	at 3s.	do.
4 other petardiers	at 2s.	do.

All of them to be nominated, constituted and appointed from time to time by the Master-General of our Ordnance, and to be under the

---

1 Grenado shells, from *granata*, a pomegranate, which this missile originally resembled, being filled with broken glass, similar to the stones or grains of the fruit.

2 Here we have the original institution of the now familiar office of "Bombardier." It is expressly stated that the bombardier was one specially skilled in the use of mortar pieces, and when mortar pieces were embarked in bomb-vessels, or formed a part of any "train," they were always accompanied by a proportion of bombardiers, who served under the orders of the fire-master and the fire-workers. They subsequently also performed the duties of the petardiers under the chief petardier, who was an officer of superior grade, as will appear by inspection of the several rates of pay given in the detail of the train raised by warrant of the 28 February, 1691 (page 123), when the office was held by Captain William Wood.

The firemaster, who received ten shillings a day as pay, was apparently the chief "Artillery Officer." (For a detail of his duties see page 77 *ante*.)

direction and guidance of our Master Fire-master, strictly ordering and commanding that in the choice of the said bombardiers all dilligent circumspection be used, and that you suffer none to be admitted to the said employments, but such only as you shall be really assured are of sufficient knowledge and abilities for that our service : and further our will and pleasure is, that the said number of bombardiers and petardiers before mentioned be continued on the quarter books of our said office, and receive out of the treasury thereof, quarterly, the several pays, allowances and entertainments hereby to each of them made and established during our royal pleasure.

By His Majesty's command,
SUNDERLAND.

To the Lord Dartmouth, Master-General of the Ordnance, or to the Master-General of the Ordnance for the time being.

1686. Dec. Quarter-books:

1 chief bombardier, Daniel Moody	Paid for the three months.
12 bombardiers	
1 chief petardier, John Fawcett	
4 petardiers	
60 gunners	

Mr. Jacob Richards was present at the siege and taking of Breda, by the imperial army, under the conduct of the Duke of Lorraine, by order of Lord Dartmouth, master-general of the ordnance, in order to his improvement for his majesty's future service.—*Harl. MS.* 4989.

1687. 22 Mar. His majesty's warrant, directing that instead of the old establishment of the office of ordnance and train of artillery in Ireland, the following should take place from this date :—

	£	s.	d.
Master of the ordnance	500	0	0
Lieutenant of the ordnance	300	0	0
Surveyor or comptroller	200	0	0
Clerk of the ordnance and deliveries	100	0	0
Storekeeper	100	0	0
Chief engineer	200	0	0
Master-gunner	50	0	0
His mate	25	0	0
Armourer	25	0	0
30 other gunners at 9d. a day each	410	12	6
6 clerks of stores at £28 per annum each	168	0	0
1 more at £26 5s.	26	5	0
1 gunner at Dungannon	16	16	0
1 do. Cork	16	16	0
1 do. Kinsale	16	16	0
1 do. Charles Fort	16	16	0
1 do. mate there	14	0	0
1 do. Limerick	16	16	0

		£	s.	d.
1 gunner at Galloway		16	16	0
1 do.	Athlone	16	16	0
1 do.	Londonderry	16	16	0
1 do.	Carrickfergus	16	16	0
1 do.	mate, Culmore	14	0	0
1 do.	do. Passage	14	0	0
1 do.	do. Waterford	14	0	0

1687. 29 Mar. His majesty's warrant: WHEREAS we think it requisite for the good of our service that some fitting person should be sent into foreign parts to improve himself in the art military, and to survey, learn and observe the best manner and method of fortifications and modern practice of artillery; and whereas you have recommended to us our trusty and well-beloved Edward Clarke, Gentleman, as a person fitly qualified for that employment: our will and pleasure is, and we do hereby authorise and require you to give immediate directions to him, the said Edward Clarke, with all convenient speed to take a journey to Hungary, and repair to the Emperor's army or armies this next campaign, and there to observe and take notice of their method of marching, encamping, embattling, exercising, ordering of their trains of artillery, their manner of approaching, besieging, or attacking any town, their mines, batteries, lines of circumvallation and contravallation, their way of fortification, their foundrys, instruments of war and engines, or what else may occur observable in his journey thither, being upon the place, or in his return back, and to make a true and perfect representation thereof unto us. And for his encouragement herein, our will and pleasure is that you allow him the salary of twenty shillings per diem during his absence upon this service, and further to make unto him by way of imprest such advance of monies as you shall judge reasonable for his better accommodation in this journey. (See page 111.)

5 June. His majesty's warrant, directing a train of artillery, with stores, ammunition, officers, gunners, bombardiers, and other ministers to be sent to our camp at Hounslow,[1] viz.:—

Brass ordnance.	2 demi-culverins	of 12 feet long.
	4 minions	„ 10 „
	6 three-pounders	„ 6 „
Do.	2 three-pounders	...for the Queen's Regiment.
	2 do.	...Princess's Regiment.
	2 do.	...Col. Cornwallis's Regiment.
	2 falconetts	...Sir E. Butes' Regiment.
	2 do.	...Col. Fulton's Regiment.
	2 three-pounders	...Col. Buchan's Battalion.

Also 6 brass mortar-pieces of 7¼ inch diameter, on travelling carriages.[2]

---

[1] This was in addition to the guns already in camp, as detailed on page 105, where we find seven other regiments in possession of two regimental guns each, making a total of 26 regimental guns and 18 guns of the artillery "train."

[2] See Macaulay's History of England, chap. vi. These, with the guns already detailed,

1688. June Quarter-book. George Brown appointed chief bombardier. 60 gunners paid quarterly.

27 Oct. From the Tower, armour lent to Colonel William Legg, for his own use, by order of the board of ordnance:

Harquebuss Amr. carbine proof	Backs......... 2	Pott 1
	Breasts 2	Gauntlet...... 1

30 Oct. To Jacob Richards, for his own use in the train of artillery by order of the Board:—

Silk armour, 1 suit.

On the 1st November, 1688, the Prince of Orange landed in England, and a train of 26 pieces was immediately organized to "attend his Majesty's forces, consisting of 15,000 foot and 5000 horse." The warrant for this train bears date, 8th Nov. 1688, although a preparatory warrant was issued in the previous month, as follows:—

1688. 15 Oct. "JAMES R.

"Right trusty and well-beloved, we greet you well. Forasmuch as we have received undoubted advice that a great and sudden invasion with an armed force of foreigners will speedily be made in an hostile manner upon this our kingdom, we find it necessary at this juncture of time to have in readiness a marching train of artillery with all other necessary stores and equipage thereunto belonging, our will and pleasure is, and we do hereby authorize and require you to provide and to cause to be made ready for our service, a train of artillery of twenty pieces of ordnance and six field mortars with all necessary stores and equipage thereunto belonging, to march upon orders sent from us or any of our Lieut.-Generals appointed to command our forces; and our further will and pleasure is, that the list presented by you unto us and hereunto annexed of the officers and ministers fitting to attend the said train if we go in person (which by us is approved of), be established according

---

made a total at the camp in 1687, of 26 regimental guns, 6 mortars, and 12 heavy guns. To these were attached the usual stores, ammunition, officers, gunners, bombardiers, and other ministers. The camp, which was first formed in May, 1685, for the purpose of overawing the city, failed in this object, but was on the contrary a great source of amusement to the town, and so continued, until at last the sentiments of the army were exactly the same as those of the populace against whom it was intended to act.

"The general satisfaction with which the famous acquittal of the bishops was received, was more than echoed by the soldiery at Hounslow, where the king received the news, and the camp was accordingly broken up in 1688, and the troops sent to quarters in different parts of the country."

Lieut.-General. Comptroller-General. Principal Engineer. Master-gunner of England, and their clerks. Chief Firemaster, and mate. Keepers, and makers of our Royal tents, and their assistants.

to the several pays per diem proposed, but if we should not think fit to march with our army, the attendance of the officers mentioned in the margin will not be necessary; and that out of such monies as are and shall be issued to the use and service of our ordnance, you cause such sums of money to be advanced as you shall judge reasonable for the present towards the defraying of the charge of the said train upon the expedition commanded, and also from time to time you cause such other sum or sums of money as shall be required for the defraying of the pays of the several officers and ministers, and carrying on our said service, to be duly made, allowed, and distributed, and for so doing this shall be your warrant.

"Given at our Court at Whitehall, this 15th day of October, 1688,

"By His Majesty's command,

"MIDDLETON."

The mattrosses on this service were clothed as under—

Striped jackets and breeches,
Blue stockings,
Leather caps.

The pioneers—

Red cloth coats,	Woollen stocking,
Red kersey breeches,	Shoes and buckles,
Leather Montreur caps,	Neckcloths.

Conductors—Red cloth cloaks.

The chief engineer—A suit of silk armour.

The mattrosses were armed with half-pikes and hangers.

1688. 8 Nov. His majesty's warrant for issuing out of H. M. stores within the office of the ordnance, a train of artillery consisting of 26 pieces of ordnance, with their carriages, stores, habiliaments of war, with all equipage necessary to attend his majesty's forces.

Nature of Ordnance.

	Number.
9 pounders	10
$5\frac{1}{4}$ do.	8
4 do.	4
3 do.	4

8 Nov. A list of the proper officers, ministers, and attendants of the train of artillery, with their names and offices:—

		Pay per diem. £	s.	d.
Sir Henry Shire	Lieutenant-general	3	0	0
Mr. John Blake	His secretary	0	5	0
	A clerk under him	0	4	0
	Comptroller-general	2	0	0
	His two clerks, at 4s. each	0	8	0

		Pay per diem. £	s.	d.
Captain John Lewin	Comptroller to the train	0	15	0
	His clerk	0	4	0
Mr. Abell Barton	Commissary of the train and ammunition for the army	0	10	0
	His two clerks, at 3s. each	0	6	0
Mr. Willm. Willougby	Adjutant of the train	0	10	0
Mr. Edward Hubbald	Paymaster	0	8	0
	His clerk	0	4	0
Mr. Edward Clark[1]	Comptroller of the baggage train	0	10	0
	His clerk	0	2	0
His Majesty's	Chief engineer	1	0	0
	His clerk	0	4	0
Mr. Jacob Richards	Engineer	0	10	0
	His clerk	0	4	0
Capt. Just. Sherburne	Waggon-master	0	10	0
	His assistant	0	4	0
Mr. Thomas Hardyman	Commissary of the draught horses	0	8	0
	His assistant	0	4	0
David Heard Francis Rainsford Obed. Beal Arskin Walker Joseph Eden Mr. Spurway	Gentlemen of the Ordnance, at 5s. each	1	10	0
Mr. Robert Barker	Quarter-master	0	5	0
Sir John Warren	Purveyor	0	5	0
Christopher Sympson	Provost marshall	0	6	0
George Raigall Thos. Harwood	His assistants, at 2s. 6d. each	0	5	0
Mr. Woolferman	Chief firemaster	0	8	0
Mr. Nelson	His mate	0	6	0
George Brown	Firemaster to the train	0	5	0
John Baxter	His mate	0	4	0
4 assistants to the firemaster	at 3s. each	0	12	0
George Brown	Chief bombardier	0	3	0
	12 bombardiers, at 2s. each	1	4	0
	4 petardiers, at 2s. each	0	8	0
Captain Leake	Master-gunner of England	0	13	4
John Fawcett	Master-gunner of the train	0	5	0
	His two mates, at 3s. each	0	6	0
	30 gunners, at 2s. each	3	0	0
	52 mattrosses, at 1s. 6d. each	3	18	0
Captain Allured	Battery-master	0	10	0
	His two assistants, at 4s. each	0	8	0
	Bridge-master	0	8	0
	His six assistants, at 3s. 6d	1	1	0
	Tinman	0	3	6
	Chief conductor	0	5	0
	21 conductors to the train and army, at 2s. 6d. each	2	12	6

1 See page 108.

		Pay per diem.		
		£	s.	d.
Thomas Scale	Surgeon	0	4	0
Wm. Wynn	His mate	0	2	6
His Majesty's	Tent-keepers and tent-makers			
	Their two assistants, at 3s. each	0	6	0
Willm. Chamberlin	Tent-maker and tent-keeper	0	4	0
	His assistant	0	2	6
Messenger to attend the	principal officers of the train	0	4	0
	Ladle-maker	0	4	0
	Master-smith	0	4	0
	His two servants, at 2s. each*	0	4	0
	Master-farrier	0	4	0
	His four servants, at 2s. each*	0	8	0
	Master-carpenter	0	4	0
	His three servants, at 3s. each*	0	9	0
	Master-wheelwright	0	4	0
	His 12 servants, at 2s. 6d. each*	1	10	0
	Master collar-maker	0	4	0
	His three servants, at 2s. each*	0	6	0
	Master-cooper	0	6	6
	His servant*	0	2	6
	Master-gunsmith	0	4	0
	His servant*	0	2	6
Holdcroft Blood	Captain of the pioneers	0	5	0
	60 pioneers, at 1s. each	3	0	0
	2 sergeants, at 2s. each	0	4	0
	2 corporals, at 1s. 6d. each	0	3	0
	2 drummers, at 1s. 6d.	0	3	0

* Servants probably means apprentices or assistants.

Abstract of moneys necessary for putting the train in a condition for marching on the present expedition, exclusive of the charge of the stores, equipage, and ammunition for the train :—

	Per diem.			Per mensem.		
	£	s.	d.	£	s.	d.
To the pay of the officers of the train	38	8	10	1,076	7	4
The hire for 1500 horses	150	0	0	4,200	0	0
For meat for 1500 horses	112	10	0	3,150	0	0
The pay of 532 drivers	39	18	0	1,117	4	0
Incident charges, per est.	...			60	0	0
	340	16	10	9,603	11	4
The charge of 60 country waggons	...			600	0	0
				10,203	11	4

1688. 1 Dec. To Sir Martin Beckman, lent him for his own use, by order of the Board :—

Silk armour, 1 suit.

WILLIAM III. AND MARY.

1689. His majesty's warrant:

8 Mar. Right trusty and right entirely beloved Cousin and Councellor, We greet you well. Whereas the late King Charles and King James did make an establishment of the Office of the Ordnance, in a book intituled "Instructions for the Government of the Office of our Ordnance," under our Master-General thereof, committed to five Principal Officers, and signed by the said late King Charles and King James and the Principal Secretary of State, our will and pleasure and express command is, that all orders, warrants, instructions, or directions therein contained and given by the late King Charles and King James as aforesaid, shall continue in force, to be obeyed and observed by all and every person, or persons, who in our said office are or shall be concerned, as fully and amply as the same was or ought to be observed in the reign of the late King Charles and King James, and for your so doing this shall be your sufficient warrant, untill our further pleasure be known. Given at our Court of Whitehall, this eighth day of March, 1688-9, in the first year of our reign.

By his Majesty's command,
SHREWSBURY.

To Frederick Duke of Schomberg,
Master-General of our Ordnance.

[King James II., having fled the kingdom on 23rd Dec., 1688, landed at Kinsale in Ireland with a French force of about 5000 men in March, 1689. Consequently, on the 1st June, a warrant was issued for a train to be prepared for shipment to Ireland. This train was further reinforced in July by guns, &c., sent by land to Chester.]

1689.
6 Mar. His majesty's warrant to Sir Henry Goodrick, lieutenant-general of the ordnance, and to the rest of the principal officers of the same, to have in readiness for the present expedition to Ireland brass ordnance, ammunition, carriages, equipage, and other provisions of war, as follows:—

Brass ordnance mounted on travelling carriages, with limbers complete	Culverins	4
	12-pounders ...	4
	Sakers	9
Brass mortar pieces mounted on the bodies of carriages	18½-inch	1
	14¾ do.	2

An additional proportion of ordnance and other stores demanded by his Grace the Duke of Schomberg:—

Brass ordnance mounted on travelling carriages, complete	Culverins	6

With spare carriages, stores, ammunition, harness, and cloths for pioneers, as follows:—

Large blue coats, lined with orange-colour	40
Orange-coloured waistcoats	40
Blue breeches	40 pair
Stockings	40 pair

Caps, embroidered with shovel in the front	40
For two sergeants	2
For two drummers	2
For a corporal	1
Drivers	200
Cloaths for the drivers' suits	200
Oats, beans, and hay for the train horses, being in No. ...	500
Master-smith and his assistants	14

1689. 14 Mar. Account of the charges laid before the House of Commons for the artillery, hospital, &c., in the late expedition to England:—

£85,612 19*s.* 8*d.*

20 Mar. Sir Robert Howard delivered to the House an estimate of the expense of the late King James, which, on a medium of four years, was as follows:—

Ordnance, £83,493 9*s.* 3*d.*

21 Mar. Estimate for the Ordnance delivered to the House:—

Ordnance, £81,935 17*s.* 3*d.*

Further supply for providing artillery and other necessaries for the office of ordnance.

£81,935 17*s.* 3*d.*

His majesty's warrant directing that several quantities of broken iron ordnance and shot now remaining in our stores at Woolwich, and several granade shells at our laboratory at Greenwich, be recast into ordnance.

12 April. His majesty's warrant for transporting two brass ordnance and two mortar pieces, with their ammunition, stores, and equipage, to Scotland, and on their arrival at Leith to be delivered to our trusty and well-beloved Major-General Mackay.

24-pounders		2
Brass mortars	11-inch	1
	10-inch	1

28 May. His majesty's warrant to Frederick Duke of Schomberg, master-general of the ordnance, for a train of artillery of brass ordnance be forthwith provided and sent to Ireland, with their stores, equipage, officers, and ministers, with their pay, &c.

	Per diem. £	s.	d.
To his Grace the Master-general...	10	0	0
His secretary or chief clerk	1	0	0
His clerk	0	5	0
Comptroller-general	1	10	0
His clerk	0	5	0
Comptroller	0	15	0
His clerk	0	4	0
Commissary of the stores	0	10	0
His clerk	0	4	0

	Per diem. £	s.	d.
Adjutant	0	8	0
Pay-master	0	8	0
His clerk	0	4	0
Assistant	0	3	0
First engineer	0	15	0
His clerk	0	4	0
Second engineer	0	10	0
Third engineer	0	8	0
Waggon-master and commissary of the draught horses	0	8	0
Two assistants, each 4s.	0	8	0
Battery-master	0	12	0
Two assistants, each 4s.	0	8	0
Storekeeper of the provisions of victuals and provender	0	6	0
His steward and clerk	0	3	6
4 gentlemen of the ordnance, each 5s.	1	0	0
Quarter-master	0	6	0
Bridge-master	0	6	0
2 mates, each 3s. 6d.	0	7	0
18 servants, each 2s.	1	16	0
Purveyor	0	5	0
His assistant	0	3	6
Provost marshall	0	6	0
2 assistants, each 4s.	0	8	0
Surgeon	0	5	0
His mate	0	3	0
Chief conductor	0	4	0
12 conductors, each 3s.	1	16	0
Tent-maker	0	4	0
His assistant	0	2	0
Master-farrier	0	4	0
His mate	0	2	0
Master-bricklayer	0	3	0
4 assistants, each 2s. 6d.	0	10	0
Master-smith	0	4	0
4 servants, each 2s.	0	8	0
Master-carpenter	0	4	0
2 mates, each 3s.	0	6	0
15 servants, each 2s. 6d.	1	17	6
Master-wheelwright	0	4	0
6 servants, each 2s. 6d.	0	15	0
Master collar-maker	0	4	0
2 servants, each 2s. 6d.	0	5	0
Ladle-maker	0	3	0
Master-cooper	0	4	0
2 servants, each 2s. 6d.	0	5	0
Master-gunner of the train	0	6	0
His two mates, each 3s.	0	6	0
22 gunners, each 2s.	2	4	0
32 mattrosses, each 1s. 6d.	2	8	0
Fire-master	0	5	0
His mate	0	3	0

	Per diem. £	s.	d.
4 fireworkers, each 2s. 6d.	0	10	0
Tinman	0	3	0
Captain of the pioneers	0	5	0
2 sergeants, each 2s.	0	4	0
2 corporals, each 1s. 6d.	0	3	0
2 drummers, each 1s.	0	2	0
30 pioneers, each 1s.	1	10	0
Drivers and attendants upon the train, 200, each 1s. 6d.	12	10	0
For provender for 600 horses, each 1s. 6d.	45	0	0
Incident charges	6	0	0
Total	105	7	6
Total	£38,461	16	6

Arms, &c., for the Gunners and Mattrosses.

Halberts	3
Short pikes	80
Hangers with belts ...	30
Long carbines, straped	146
Cartouch boxes for do.	146
Pistols	50 pairs
Troop saddles	18
Bitt bridles	18

1689. 1 June. His Majesty's warrant directing the issue and shipment of the following train of artillery for the service of the present expedition into Ireland:—

	No.
Culverings	2
Sakers	4
3-pounders	12
18-inch mortar	1
10 do. do.	4

with their ammunition, stores, carriages, &c.

8 June. The Duke of Schomberg writes to Sir Henry Goodricke, "that he on Thursday last viewed the horses lately bought for the train, and they appear to me too slight for that service. Besides, one was lame, another an old coach-horse, and most of them faulty. His majesty intends to have 400 only.

"But I desire you will consult the best for the service, and let no horses be bought upon this occasion but what are strong and serviceable."

1 July. His majesty's warrant for a train of brass ordnance and mortars, with their stores, equipage, and our officers and ministers attending on the same, for Ireland.[1]

[1] This detail is almost identical with that given under date 28th May, and probably refers to the same train, but the names of the officers and ministers are now added.

Name	Office	Per diem. £	s.	d.
His Grace the Duke of Schonberg	Master-general	10	0	0
Willm. Holford, Esq.	his secretary or chief clerk	1	0	0
His clerk		0	5	0
Mon. De Gaulon	Lieut.-general	1	0	0
His clerk		0	5	0
Capt. Edwd. Clerke	Comptroller	0	15	0
His clerk		0	4	0
Mon. De Glumm	Comptroller of the fire-master and train	0	12	0
Robt. Aleway	Commissary of stores	0	8	0
Edwd. Gibbon	His clerk	0	4	0
Adam Blackford	Adjutant	0	6	0
Wm. Hubbald	Pay-master	0	8	0
Clement Ireson	His clerk	0	4	0
Thos. Newton	An assistant	0	3	0
Jacob Richards	First engineer	0	15	0
His clerk		0	4	0
Capt. Holcraft Blood	Second engineer	0	10	0
Claudus Barden	Third engineer	0	8	0
George Barnard	Waggon-master	0	10	0
His two assistants, each 4s.		0	8	0
Gilbert Filken	Commissary of the draught horses	0	15	0
Henry Covall	Battery-master	0	12	0
John Jackson	His assistant	0	4	0
John Read	Storekeeper of the victuals and provender	0	5	0
Thos. Hawes	His steward or clerk	0	8	0
Thos. Dilke Walter Thomas John Richards Obadiah Field Obadiah Adney John Child	Six gentlemen of the ordnance, 5s. each	1	10	0
David Heard	Quarter-master	0	6	0
Godfry Richards	Purveyor	0	5	0
His two assistants, each 3s.		0	6	0
Thos. Deane	Provost marshall	0	3	0
His two assistants, each 2s. 6d.		0	5	0
Physition		0	8	0
Chirurgeon		0	5	0
His mate		0	2	6
Thos. Knox	Chief conductor	0	5	0
14 conductors, each 3s.		2	2	0
Thos. Chamberlain	Tent-maker	0	4	0
His assistant		0	2	0
Richd. Bedford	Master-farrier	0	2	6
His mate		0	2	6
Rowland Smith	Master-smith	0	3	0
2 mates, each 2s.		0	4	0
Richard Giles	Master-carpenter	0	3	0
2 mates		0	6	0

		Per diem. £	s.	d.
15 servants, each 2s. 6d.		1	17	6
4 bricklayers, each 2s. 6d.		0	10	0
7 wheelwrights, each 2s. 6d.		0	17	6
John Mitchell	Master collar-maker	0	3	0
5 servants, each 2s. 6d.		0	12	6
Richd. Mahue	Master-cooper	0	3	0
3 servants, each 2s. 6d.		0	7	6
Robt. Berry	Ladle-maker	0	3	0
3 tinmen, 2s. 6d. each		0	7	6
Edwd. Pickering	Master-gunner	0	5	0
2 mates, each 3s.		0	6	0
40 gunners, each 2s.		4	0	0
50 mattrosses, each 1s. 6d.		3	15	0
Robt. Van Orb	Bridge-master	0	6	0
John De Bonnez	His assistant	0	3	0
10 servants, each 2s.		1	0	0
Isaac Nelson	Fire-master	0	5	0
John Hopeke	His mate	0	3	0
Andrew Shaw Jacob Walferman	Fire-workers, each 2s. 6d.	0	5	0
7 bombardiers, each 2s. 6d.		0	17	6
Andrew Hydeman	Tinman	0	3	0
John Owens	Captain of the pyoneers	0	5	0
2 corporals, each 2s. 6d.		0	5	0
3 drummers, each 1s.		0	3	0
Incident charges		6	0	0
100 carters, each 1s. 6d.		7	10	0
100 boys, each 6d.		2	10	0
6 sumpter men, each 1s.		0	6	0
For provender for 600 horses, each 1s. 6d.		45	0	0
	Total per diem.	£105	9	6

Memo.

But 492 horses are set down in the general estimate, which are only for the ordnance and stores, the rest of the addition of the 600 arrising by the pontoons, &c. The total annual allowance to the several officers, ministers, and attendants of the train, &c. ... £38,498 7 6

1689. 16 July. The Duke of Schomberg appointed Mr. George Barnard waggon-master to this office, and to attend the train of artillery designed for Ireland.

The Duke ordered that the gunners, mattrosses, and tradesmen have coats of blue, with brass buttons, and lined with orange bayes,[1] and hatts with orange silk galoone.

The carters' grey coats, lined with the same.

July. An additional proportion of brass ordnance and stores of war ordered to be dispatched by land to Chester, according to his majesty's command :—

3-pounders 3

[1] *Qy.* Baize.

1689. July. A further additional proportion sent to Chester:—

Minions 4

Aug. Colonel Richards wounded in the trenches at the siege of Carrickfergus.

12 Aug. The Duke of Schomberg with his army landed near Belfast, and marched from thence to Carrick Fergus, to which he laid siege on the 22nd: it surrendered the 26th. The army then advanced to Dundalk, where his Grace fortified himself and remained the winter.

1690. 14 June. King William landed near Belfast, and gave orders for assembling the army at Dundalk.

1 July. He fought the Battle of the Boyne, where his majesty defeated the enemy, the French passing the Shannon at Athlone, marching to Galloway, and embarked for France.

The brave Duke of Schomberg was killed in this battle.

1690. Chief fire-master, Issac Nelson. January Quarter Book.
Ditto Johanen Signum Schlundt. December Quarter Book.

[On the return to England of King William III., after the battle of the Boyne, we hear of no more levies, until the king, in the year 1691, set out for Holland with Marlborough, to take the command of the army against the French. The king did not land in Holland until June, 1691, but on the 27th of the previous February, a warrant was issued to prepare "a train of brass ordnance, with all fitting equipage, to be forthwith transported into Flanders for our service." It will be observed that the guns of this train were all light guns. On the following day, another warrant was issued for a train of artillery of 76 pieces "for this summer's expedition by sea." The train was originally intended to be composed of the following pieces, viz.,

Fourteen	24-prs.	Ten	3-prs.
Sixteen	18 „	Eight	13-inch mortars
Eight	12 „	Twelve	10 „
Six	6 „	Two	8-inch howitzers.]

1691. 27 Feb. To Sir Henry Shaw, Lieutenant-General of the late train of artillery:—

Harquebuss armour, carbine proof...	Backs ...	1	Pott	1
	Breast...	1	Elbow gauntlet ...	1

27 Feb. His majesty's warrant to Sir Henry Goodricke, Lieutenant-General of the Ordnance, to have in readiness a train of brass

ordnance, with all fitting equipage, to be forthwith transported into Flanders for our service, viz. :—

Demi-culverins	8
Sakers	10
Three-pounders	20
Howitzers	4
Petards (small)	2

with the following officers, &c. :—

	£	s.	d.
Colonel John Wynant Goor and his clerk	2	0	0
Comptroller and his clerk	3	0	0
Lieutenant-Colonel Jacob Richards and his clerk	0	19	0
Major John Simon Schlunt	0	16	0
Captain Lieutenant	0	11	0
Battery-master	0	12	0
His assistant	0	4	0
Adjutant	0	8	0
Quarter-master	0	8	0
Chaplain	0	8	0
Auditor	0	6	0
Pay-master	0	8	0
His assistant	0	4	0
Master-surgeon and his assistant	0	10	0
6 gentlemen of the ordnance, at 7s. 4d. each	2	4	0
4 engineers, at 10s. each	2	0	0
Petardier	0	6	0
4 fire-masters, at 5s. each	1	0	0
4 bombardiers, at 2s. 6d. each	0	10	0
2 commissaries of stores, at 8s. each	0	16	0
2 clerks of stores, at 4s. each	0	8	0
8 conductors of stores, at 3s. each	1	4	0
Conductor plumber, with his assistant	0	4	0
11 conductors, at 3s. each	1	13	0
Captain of the tin-boats	0	6	2
His assistant	0	3	0
2 corporals, at 2s. 6d. each	0	5	0
40 private men, at 2s. each	4	0	0
Master-tinman	0	5	0
2 assistants, at 3s. each	0	6	0
Master-gunner	0	5	0
2 mates, at 3s. each	0	6	0
2 corporals, at 2s. 6d. each	0	5	0
40 gunners, at 2s. each	4	0	0
Sergeant-miner	0	2	6
9 miners, at 1s. 6d. each	0	13	6
Waggon-master	0	10	0
2 assistants, at 4s. each	0	8	0
Captain of the carpenters	0	6	3
2 mates, at 3s. each	0	6	0
18 carpenters, at 2s. 6d. each	2	5	0
Master-wheelright	0	4	0
6 wheelrights, at 2s. 6d. each	0	15	0

	£	s.	d.
Master-smith	0	4	0
4 smiths, at 2s. 6d. each	0	10	0
Master collar-maker	0	4	0
6 collar-makers, at 2s. 6d. each	0	15	0
2 coopers, at 2s. 6d. each	0	5	0
Tent-maker	0	4	0
Kettle drummer [1]	0	4	0
His coachman [1]	0	3	0
Provost marshall	0	3	0
2 assistants, at 2s. 6d. each	0	5	0
80 mattrosses, at 1s. 6d. each	6	0	0
	45	7	5
Hiring 600 horses for the guns and tin-boats	75	0	0
For 200 waggons, with 600 horses, for stores	80	0	0
Contingent charges	20	0	0
Per diem	220	7	5
Per week	1,542	11	11
Per month	6,170	7	8
For seven months	43,192	13	8
Value of stores, &c.	27,614	7	1
	£70,807	0	9

1691. 28 Feb. His Majesty's warrant to Sir Henry Goodricke, directing a train of brass ordnance, mortars, &c., together with a fitting proportion of ammunition and other stores, to be in readiness to be put on board such ships as shall be provided, to be employed in such service as we shall think fit.[2]

Number and Nature of Ordnance.

24-pounders	14
18 do.	16
12 do.	8

1 "Kettle drummer and his coachman." This is the first mention of the great kettle drums which accompanied the train of artillery throughout Marlborough's campaigns, and formed a conspicuous feature at Marlborough's funeral. They are represented in the model of Marlborough's train at the R.M. Repository, and the silk and gold-embossed housings are now (April 1880) in the Harness Store at the Royal Arsenal. I have not been able to trace the kettle drums themselves.

2 It would appear from stray notes in the MS., that of this detail eight 13-inch mortars only were at once despatched to Ghent. A subsequent warrant, dated 6 May, 1693, orders this train, slightly altered in detail, to be actually "SHIPPED." As every change in the detail of a train appears to have required a new warrant to be issued for the whole, there is some confusion, but the total number of guns under Sir Martin Beckman would appear to have been 86. We find the British artillery engaged in all the actions and sieges of this period, including the battles of Steinkirk (1692), Landen (1693), seiges of Namur (1692–95), and Huy (1694).

6-pounders	6
3 do.	10
13-inch mortars	8
10 do. do.	12
8 do. howitzers	2

List of officers, &c., appointed to this train, with their pay per diem:—

		£	s.	d.
Colonel and clerk	Col. Sir Martin Beckman	2	0	0
Comptroller and his clerk		1	0	0
Lieut.-colonel and his clerk	Lieut.-Col. George Brown	0	19	0
Major	John Henry Hopeke	0	15	0
Captain-lieutenant	Leod. Vanderstam	0	10	0
Battery-master	Anthony De Yong	0	12	0
His assistant	Janett Cook	0	4	0
Adjutant	Claude Testfoile	0	8	0
Quarter-master	Matthew Purcelle	0	8	0
Chaplain	Wm. Gallway	0	8	0
Pay-master	John Jennings, Esq.	0	8	0
Assistant to the pay-master	George Spencer	0	4	0
Master-surgeon	William Clark	0	8	0
2 assistants, at 3s. each	James Benoist Fras. Tompkins	0	6	0
8 gentlemen of ordnance at 7s. 4d. each	Legg Goodricke Dilke Harang Wheeler Holeman Shelden Gibbins	2	18	8
12 engineers at 10s. each		6	0	0
Petardier	Captain William Wood	0	6	0
His assistant	William Wood	0	2	6
Fire-master	Captain Thomas Browne	0	10	0
19 fireworkers at 5s. each		4	15	0
30 bombadiers at 2s. 6d. each		3	15	0
2 commissary of stores at 8s. each	Samuel Palling David Lewis John Silvester	0	16	0
2 clerks of stores at 4s. each	Lewis Hardwicke Thomas Rushell	0	8	0
18 conductors at 3s. each		2	14	0
Captain of the tin-boats	Gavert Van Erp	0	6	0
An assistant	William Bridgman	0	3	0
2 corporals at 2s. 6d. each		0	5	0
30 bridge-men at 2s. each		3	0	0
Master-tinman	George Faucett	0	5	0
2 men at 3s. each		0	6	0
Master-gunner of England		0	13	4
92 gunners at 2s. each		9	4	0
92 mattrosses at 1s. 6d. each		6	18	0
Waggon-master	William Barnes	0	10	0

		£	s.	d.
His assistant	John Wayde	0	4	0
Master-carpenter	John Cooper	0	4	0
19 carpenters at 2s. 6d.		2	7	6
Master-wheelwright	Thomas Bortman	0	4	0
6 wheelwrights at 2s. 6d.		0	15	0
Master-smith	Christopher Dixon	0	4	0
7 smiths at 2s. 6d. each		0	17	6
Master-farrier	Richard Marshall	0	4	0
2 farriers at 2s. 6d. each		0	5	0
Master collar-maker	Roger Tupper	0	4	0
2 men at 2s. 6d. each		0	5	0
Master-cooper	William James	0	4	0
Assistant		0	2	0
Tent-maker	John Gilbert	0	4	0
Assistant		0	3	0
Basket-maker and his assistant		0	5	0
Provost marshall	Thomas Glendall	0	3	0
2 assistants at 2s. 6d. each		0	5	0
2 conductors of woolpacks at 8s. each		0	16	0
For keeping 200 horses at 1s. 3d. each		12	10	0
70 drivers at 1s. 6d. each		5	5	0
4 conductors of horses at 3s. each		0	12	0
Purveyor	Capt. Godfry Richards	0	8	0
His assistant		0	3	0
Commissary of the draught horses	Daniel Sherrard	0	8	0
His assistant	Thomas Hall	0	4	0
Contingent charges		20	0	0
	Per diem	99	13	6
	For 6 months	16,745	8	0

	£	s.	d.
Pay	16,745	8	0
For ordnance and stores	102,587	14	4
Total	£119,333	2	4

1691. May. General Ginkle besieged Athlone: the artillery had 50 pieces of cannon, besides mortars, which battered down the face of a tower and levelled the out trench along the river; they also kept up a constant fire on the enemy's works over the heads of the storming party while crossing the river to the attack.

12 July. At the battle of Aghrim, fought between General St. Ruth, commanding the Irish army, and General Ginckle, Captain Logan of the artillery with one of his field pieces took off the head of General St. Ruth.

Sept. Colonel Goor (with the British artillery) commanded in the trenches, with the allied army, at the siege and surrender of Mentz.

In the Quarter-books: bombardiers, petardiers and gunners, paid as the previous year.

1692. Albert Borgarde entered and joined the English artillery in Flanders at the Camp of Mille as Fire-master under the command of Colonel Goor, and was present with the artillery at the battle of Steenkerke, and afterwards joined Sir Martin Beckman at Ostend, who had arrived from England with a detachment of artillery.

6 March. Right trusty and well beloved we greet you well. WHEREAS you have represented to us that by reason of the great scarcity of field gunners at this juncture to serve us abroad it is necessary for our service that draughts of experienced gunners be drawn out of our several garrisons, castles, and fortifications in England to serve us abroad, and their vacancies, if our service shall so require, to be supplied by practioner gunners, our will and pleasure therefore is, and we do hereby authorize and empower and direct you from time to time to cause such draughts of field gunners to be made accordingly for our service abroad, and when you shall find it necessary for our service you cause their vacancy to be supplied by practioner gunners as afforesaid, and for so doing this shall be as well to you as to all and every of our governors, lieutenant-governors or deputy governors of our said garrisons, castles, and fortifications, a sufficient warrant. Given at our court at Whitehall, this 6th day of March, 1692, in the fifth year of our reign.

By His Majesty's command,
NOTTINGHAM.

To Sir Henry Goodricke, &c.

7 April. Her majesty's warrant, that you cause twelve cashé pieces of brass of seven foot in length and ten inches in diameter, and likewise a quantity of strong powder according to the late invention of Sir Pollycarper Wharton, to be forthwith made and provided for our service, &c.

(Signed) NOTTINGHAM.

To Sir Henry Goodricke, &c.

28 May. To Abrah. Butler, chief petardier of the Flanders train, per warrant from his majesty:—

Harquebuss armour, musquet proof	Back	1
	Breast	1
	Pott	1

30 June. Colonel Wolfgang William Romer appointed 1st engineer of our train of artillery for the present expedition by sea.

4 Aug. Her majesty's warrant to Sir Henry Goodrick, Kt. and Bart., lieut.-general of the ordnance, to send brass ordnance and mortars, with their equipage, stores, and habiliaments of war, with engineers, bombardiers, gunners and mattrosses, to the West Indias.

		Ordnance.	No.
Brass	Travelling carriages	Sakers	6
	Bodies of carriages	11 inches	2
	—	Petards	4

	Per diem. £	s.	d.
First engineer	0	10	0
Second engineer	0	6	0

	Per diem. £	s.	d.
2 bombadiers, each 3s.	0	6	0
3 miners, each 2s. 6d.	0	7	6
Master-gunner	0	6	0
His mate	0	3	0
5 gunners, each 2s. 6d.	0	12	6
6 mattrosses, each 2s.	0	12	0
Total per diem ...	3	3	0
At that rate for 183 days amounts to ...	£576	9	0

1692. 10 Sept. Captain Thomas Browne appointed by his majesty's warrant, chief fire-master to the train of artillery provided for service in the West Indias, with an allowance of 10s. per diem.

4 Oct. Sent to the West Indias for 2 engineers by Board's order:—

Harquebuss armour, musquet proof	Breast	2
	Back	2
	Potts	2

1693. 14 Feb. PERQUISITE[1] allowed by King William the Third to the Commanding Officer of Artillery in Ireland, as is shown by the following warrant:—

William R.: Whereas by our royal warrant bearing date the 25th day of February, in the 4th year of our reign, we did authorize and empower the Lieutenant-General and Principal Officers of our Ordnance to pay to our trusty and well-beloved Colonel John Whynant Goor, the sum of £500, in consideration of several broken and unservicable brass ordnance, &c., found in the towns reduced during the war in our kingdom in Ireland, being a perquisite belonging and apertaining to the said Colonel Goor, and were by him delivered into our magazines for our future service, and whereas the said Colonel Goor hath not yet received any part of the £500, by reason it was to be paid out of such monies as should be appointed for payment of the arrears of the train in Ireland, which payment we have not yet thought fit to direct; we are therefore, out of our royal favour to the said Colonel Goor, graciously pleased to direct you to cause the said sum of £500 to be paid out of the monies appropriated to the Office of Ordnance on account of land service, and for so doing this shall be as well to you as to the auditor of our inquests, a sufficient warrant. Given at our court at Whitehall, this 14th day of February, 1693, in the sixth year of our reign.

By His Majesty's command,

J. TRENCHARD.

To our right trusty and well-beloved cousin and counseller, Henry Viscount Sidney, Master-General of our Ordnance.

MEM.—The Master-General of the Ordnance, his signification upon the above said warrant, dated 14th day of February, 1693.

---

[1] See page 22 *ante* for perquisite of the church bells of a captured town, as granted to the Artillery.

1693. 5 Mar. His majesty's warrant to Henry Viscount Sidney, master-general, directing an additional train of brass ordnance, ammunition, stores, with officers, gunners, &c., for our service in Flanders.

Nature of Ordnance provided.

Demi-culverins	10
Sakers	30
3 pounders	20
8-inch howitzers	4

6 Mar. WHEREAS his majesty, by his warrant under his royal sign manual, dated the 5th day of March, 1693, hath authorised and required me to cause a train of brass ordnance and equipage thereunto belonging to be provided and transported into Flanders for his service there, and that the officers, gunners, and attendants of the said train should be regulated and established according to their respective pays per diem, as mentioned in a list annexed to the said warrant, and I being satisfied as well of the ability and sufficiency as of the loyalty and fidelity of you, the said John Wynant Goor, do hereby nominate, constitute, and appoint you, the said John Wynant Goor, to be Colonel of the said train of artillery. You are, therefore, to take the said train of artillery into your care and charge, as Colonel thereof, and to use your best endeavours to keep the said train in good order and discipline, and you are to observe and follow such orders and directions as you shall from time to time receive from his majesty, myself, or any other, your superior officers, according to the rules and discipline of war, in pursuance of the trust hereby reposed in you; and for your care and pains to be taken herein you are to receive from the paymaster of the said train, for yourself and clerk, the sum of forty shillings per diem. Given at the Office of Ordnance, under my hand and seal, this 6th day of March, 1693.

Signed, SYDNEY.

To Colonel John Wynant Goor.

Similar warrants for every individual ordered on the same service.

List of officers, ministers, &c., made choice of to attend their majesties' train of artillery in Flanders:—

	Per diem. £	s.	d.
Colonel John Wynant Goor and his clerk	2	0	0
William Musters, comptroller, and his clerk	3	0	0
Lieut.-Col. Jacob Richards	1	5	0
Major John Rymond Schlundt	0	16	0
Adjutant Daniel Cottin	0	8	0
Quarter-master Ralph Wood	0	8	0
Pay-master	0	12	0
His two assistants: John Silvister	0	3	6
His two assistants: Edmond Williamson	0	3	6
Waggon-master, Charles Ball	0	10	0
His two assistants: Godfrey Frank	0	4	0
His two assistants: Francis Ellis	0	4	0
Auditor, Charles Watkins	0	6	0
Commissary of horses, Daniel Coenen	0	8	0
Chaplain, John Carpenter	0	8	0

		Per diem. £	s.	d.
Surgeon, John De Quavre		0	8	0
His three assistants	John Grele	0	3	4
	Robert Shemer	0	3	4
	———	0	3	4
Provost, John Hill		0	3	0
His three assistants	Mathew Browne	0	2	6
	James Gold	0	2	6
	Charles Hartley	0	2	6
Kettle drummer, John Burnett		0	4	0
His coachman, John Humphreys		0	3	0

REGIMENT CONSISTING OF FOUR COMPANIES IN FLANDERS 1693.[1]

1st Company.

	Pay per diem. £	s.	d.
Captain, George Leslie	0	12	0
Lieutenants:—			
1st, George Spencer	0	8	0
2nd, Andrew Bonnett	0	7	6
Gentlemen of the Ordnance:—			
1st, Edward Grover	0	5	0
2nd, Thomas Rushall	0	5	0
4 sergeants of the gunners, at 3s. each	0	12	0
36 gunners, at 2s. each	3	12	0
4 corporals of the mattrosses, at 2s. 6d. each	0	10	0
56 mattrosses, at 1s. 6d. each	4	4	0

2nd Company.

	£	s.	d.
Captain, Abraham Decock	0	12	0
Lieutenants:—			
1st, Daniel De Young	0	8	0
2nd, Peter Gelmuyden	0	7	6
Gentlemen of the Ordnance:—			
1st, Joseph Durden	0	5	0
2nd, Samuel North	0	5	0
4 sergeants of the gunners, at 3s.	0	12	0
36 gunners, at 2s. each	3	12	0
4 corporals of the mattrosses, at 2s. 6d. each	0	10	0
56 mattrosses, at 1s. 6d. each	4	4	0

3rd Company.

	Pay per diem. £	s.	d.
Captain, Jonas Watson	0	12	0
Lieutenants:—			
1st, John George Buttensten	0	8	0
2nd, William Bousfield	0	7	6
Gentlemen of the Ordnance:—			
1st, John Hillibon	0	5	0
2nd, John Ganson	0	5	0
4 sergeants of the gunners, at 3s. each	0	12	0
36 gunners, at 2s. each	3	12	0
4 corporals of the mattrosses, at 2s. 6d. each	0	10	0
56 mattrosses, at 1s. 6d. each	4	4	0

4th Company.

	£	s.	d.
Captain, Leonard Vanderstam	0	12	0
Lieutenants:—			
1st, Albert Briluis	0	8	0
2nd, Abraham Butler	0	7	6
Gentlemen of the Ordnance:—			
1st, Joseph Lamotte	0	5	0
2nd, William Gunn	0	5	0
4 sergeants of the gunners, at 3s. each	0	12	0
36 gunners, at 2s. each	3	12	0
4 corporals of the mattrosses, at 2s. 6d. each	0	10	0
56 mattrosses, at 1s. 6d. each	4	4	0

1 The enumeration of the foregoing trains will have given some idea as to method of providing the various expeditions with artillery; but the opportunity now presents itself for noticing the first regimental train of artillery. Though it is very doubtful whether at this period the artillery had any *regimental* organisation, yet in the year 1693 the Flanders establishment consisted of 4 *companies*, and a company of "tin-boat men," or bridgemen, and a detachment of pioneers. The annexed detail is that of the *companies* then in Flanders. A reference to the details of the "trains" then in Flanders, will show a total of 137 gunners. This number divided among 4 companies will give 34 to each, which is nearly the number mentioned as the strength in each company. A comparison of the names of the officers will tend to show that the personnel of the companies included the members of the previous "trains."

	Pay per diem. £	s.	d.
Bridgemen or Tin-boat men.			
Captains:—			
1st, William Van Erp	0	6	2
2nd, Thomas Glendall	0	6	2
Lieutenant Thomas Morrice	0	4	0
4 corporals, at 2s. 6d. each	0	10	0
50 private men, ten included to be English watermen, at 2s. each	5	0	0
Master-tinman	0	5	0
2 assistants, at 3s. each	0	6	0
5 engineers, at 10s. each:— John Bodt, John Manclere, Isaac Cock, Michael Richards, Daniel Shevrard	2	10	0
Chief fire-master, John Doling	0	5	0
Fire-master and petardier, John Lewis Schlundt	0	6	0
10 firemasters at 5s. each:— John Chanternell, John Abrecht Churdes, Albert Borgard, Olaus Hulk, Henryck Lunenburg, John Suit, Erick Schiller, John Spicker, William Hendricks, John Bleckenback	2	10	0

	Pay per diem. £	s.	d.
12 bombardiers at 2s. 6d. each	1	10	0
2 commissaries at 8s. each:— Mathew Smallen, Robert Welsh	0	16	0
4 clerks of stores, at 4s. each:— Thomas Edwards, Hermaine De Weess, Thomas Fletcher, John Bourden	0	16	0
Conductor and plumber	0	4	0
28 conductors, at 3s. each	4	4	0
2 conductors and coopers, at 3s. each	0	6	0
Master-carpenter	0	6	3
2 mates at 3s. each	0	6	0
20 private men at 2s. 6d. each	2	10	0
Master-wheelwright	0	4	0
6 wheelwrights at 2s. 6d. each	0	15	0
Master-smith	0	4	0
6 smiths at 2s. 6d. each	0	15	0
Master collar-maker	0	4	0
6 Collar-makers at 2s. 6d.	0	15	0
Master tent-maker	0	4	0
Assistant	0	2	6
Pioneers.			
Lieutenant, Robert Guybons	0	4	0
4 sergeants at 2s. each	0	8	0
54 pioneers at 1s. 2d. each	3	3	0

1693. 30 Mar. His Majesty's warrant: WHEREAS by our royal warrant bearing date the 27th day of February, in the fourth year of our reign, we did authorize and require you to cause a train of brass ordnance of several natures with all necessary equipage and appurtenances to be provided for our service in Flanders: now for as much as we have thought fit to cause the brass ordnance and stores mentioned in the annexed account to be recruited and added to the said train, as also the several officers, ministers and gunners according to their respective pays per diem as mentioned in the annext list by us approved, and is hereby accordingly established, and that from time to time you cause such money to be advanced them as may defray their respective pays out of such sum or sums of money as shall be issued to you out of the Treasury on account of the train in Flanders; and whereas there is now remaining in Ghent in Flanders 8 brass mortars of 13 inches diameter with their carriages and equipage; our pleasure is, that they remain there till further orders, and our further pleasure is, that you forthwith contract with some person or persons on the best terms for our service, for providing a sufficient number of draught horses for the cannon and pontoons of the said train, as likewise for horses and waggons for the ammunition and stores during the next campaigne, or for as long as our service shall require. Our further pleasure is, that the pays of the officers, &c., mentioned in the annexed list to commence the first day of this instant March, and so continue during our pleasure, and for so doing this shall be as well to you as to all other officers concerned, a sufficient warrant. Given at our Court at White Hall, the 30th day of March, 1693, in the fifth year of our reign.

By His Majesty's command,
NOTTINGHAM.

To our right trusty and well-beloved councillor, Sir Henry Goodriche, Kt. and Bart., Lieut. of the Ordnance, and to the rest of the Principal Officers of the same.

Nature of Ordnance to recruit the Train.

6 pounders		4
3 do.		2
6 do.	chamber pieces	4

List of officers and attendants to be added to the establishment of the Flanders train of artillery, with their pay per diem :—

		£	s.	d.
2 gentlemen of the ordnance, at 7s. 4d. each		0	14	8
8 fire-workers, at 5s. each		2	0	0
8 bombardiers, at 2s. 6d.		1	0	0
2 clerks of stores, at 4s.		0	8	0
5 conductors, at 3s.		0	15	0
Smith		0	2	6
40 gunners, at 2s. each		4	0	0
Assistants to the	Pay-master	0	4	0
	Battery-master	0	4	0
	2 surgeons, at 3s.	0	6	0
	Provost-marshall	0	2	6
Bridge-men...	Lieutenant	0	4	0
	Wheelwright	0	2	6

1693. 20 April. Detach. of artillery ordered to Guernsey.

Fire-worker, Henry Brass.	20 carpenters.
3 bombardiers.	3 mason.
4 wheelers.	7 bricklayers.
3 smiths.	3 labourers.
1 collar-maker.	

6 May. Her majesty's warrant to Sir Henry Goodricke, lieutenant-general of the ordnance, directing a train of brass ordnance and mortars, ammunition, stores, and habiliaments of war, with the officers, ministers and attendants of the said train to be shipt on board such ships as may be provided.

Nature of Ordnance provided.

24 pounders	20	78
Culverins	20	
Sakers	18	
13-inch mortars	10	
10-inch do.	10	

List of proper officers and ministers necessary to attend their majesties' train of artillery, designed to sail with the fleet on this summer expedition, with their respective allowances of pay per diem:—

	£	s.	d.
Colonel Sir Martin Beckman and his clerk	2	0	0
Capt. Lumerston, comptroller, and his clerk	1	0	0
Lieut.-Colonel George Brown and his clerk	0	19	0
Major John Henry Hopkey	0	15	0
Capt. Leshley, captain-lieut.	0	10	0
Anthony De Yonge, battery-master	0	12	0
Javrett Cooke, his assistant	0	4	0
John La Predella, adjutant	0	8	0
Mathew Purcell, quarter-master	0	8	0
William Galloway, chaplain	0	8	0
George Spencer, John Steward, } assistant pay-masters, at 4s. each	0	8	0
John Bamber, surgeon	0	5	6
Francis Tomkins, 2nd do.	0	5	6
William Skeat, 3rd do.	0	3	0
8 gentlemen of the ordnance, at 7s. 4d. each	2	18	8
Captain Phillips, chief engineer	0	19	0
Engineers, 11:—			
Holcroft Blood	0	10	0
Talbert Edwards	0	10	0
Peter Carless	0	10	0
John Gobett	0	10	0
Francis Cadoul	0	10	0
Lewis Petit	0	10	0
George Conrade	0	10	0
Alexander Martinery	0	10	0
Jean Chardelou	0	10	0
Isaac Francis Petit	0	10	0
Daniel Sherrard	0	10	0
2 sub-engineers, at 5s. each	0	10	0
Captain William Wood, chief petardier	0	6	0
Thomas Taylor, his assistant	0	2	6
John Baxter, fire-master	0	10	0
Zachariah Smith, Thomas Haydon, } assistants, at 5s. each	0	10	0
17 fire-workers, at 5s. each	4	5	0
30 bombardiers, at 2s. 6d. each	3	15	0
William Barnes, waggon-master	0	10	0
James Beck, his assistant	0	4	0
2 commissaries of stores, at 8s. each	0	16	0
4 clerks of do., at 4s.	0	16	0

	£	s.	d.
2 conductors of woolpacks, at 8s. each	0	16	0
Evan Brothero, commissary of horses	0	8	0
Roger Coleburne, his assistant	0	4	0
4 conductors of horses, at 4s. each	0	16	0
John Wayle, purveyor	0	7	0
Thomas Glendall, his assistant	0	5	0
26 conductors, at 3s. each...	3	18	0
William Bridgman, master of the tin-boats	0	6	0
Richard Baker, lieut.	0	3	0
Ambrose Roberts, his assistant	0	2	6
2 corporals, at 2s. 6d. each	0	5	0
20 bridge-men, at 2s. each	2	0	0
George Fassett, master tin-man	0	5	0
James Clayton, Nicholas Curtis, } tin-men, at 3s. each...	0	6	0
Thomas Dodge, master-gunner	0	5	0
3, his mates, at 3s. each	0	9	0
100 gunners, at 2s. each	10	0	0
100 mattrosses, at 1s. 6d. each	7	10	0
2 basket makers:—			
Nicholas Benson, master	0	3	0
John Benson	0	2	0
John Cooper, master-carpenter	0	4	0
22 carpenters, at 2s. 6d. each	2	15	0
Thomas Bowman, master-wheelwright	0	4	0
11 wheelwrights, at 2s. 6d. each	1	17	6
Christopher Dixon, master-smith	0	4	0
8 smiths, at 2s. 6d. each	1	0	0
John Linsey, master-farrier	0	4	0
4 farriers, at 2s. 6d. each	0	10	0
George Tupper, master collar-maker	0	4	0
4 collar-makers, at 2s. 6d. each...	0	10	0
Nathaniel Trickpeny, master tent-maker	0	4	0
2 tent-makers, at 3s. each	0	6	0
William James, master-cooper	0	4	0
2 coopers, at 2s. 6d. each...	0	5	0
Richard Hill, provost-marshall	0	3	0
His two assistants, at 2s. 6d. each	0	5	0
10 sergeant-carters, at 2s. each	1	0	0
86 drivers, at 1s. 6d. each...	6	9	0
	74	18	8

1693. 1 July. Lord Viscount Sidney, afterwards Earl of Romney, appointed master-general.

29 July. The British artillery were in the battle of Landen.

11 Sept. Her majesty's warrant to Henry Viscount Sidney, master-general of the ordnance, to provide mortars, carcasses, grenades, shells, powder, arms, and other provisions and stores of war; with officers and attendants as specified in the annext list, to be put on

board our bomb vessels, the "Serpent," "Mortar," "Firedrake," "Grenade," and such other vessells as shall be provided.

Mortars, 13-inch 8

List of persons to attend the bomb vessels upon this expedition, with their pay per diem.[1]

		£	s.	d.
Mathew Purcell, commissary		0	8	0
John Henry Hopkey, fire-master		0	10	0
Zachariah Smith				
Thomas Hydon				
Andrew Hydeman				
John Brown				
George Warwick				
Bryan Ororch				
Henry Higgins				
Charles Kildare	fire-workers, at 5s. each ...	3	15	0
Josias Du Faux				
Joachem Miller				
John Smith				
John Winskell				
John Humphreys				
Cornelius Ambrose				
William Wood				
17 bombardiers, at 2s. 6d. each		2	2	6
Captain Thomas Silver, master-gunner		0	5	0
4 gunners, at 2s. each		0	8	0
Francis Tomkies, surgeon		0	5	0
His assistant		0	2	0
Master-carpenter		0	4	0
His mate		0	2	6
5 carpenters, at 2s. 6d. each		0	12	6
Master-cooper		0	4	0
His assistant		0	2	6
	Total	£9	1	0

1693. King William ordered a fleet under Commodore Benbow with
Nov. four bomb vessels to bombard St. Malo, viz., Serpent, Mortar, Firedrake, Grenade.

The contrivance of firing mortars from ships at sea was then a new invention, having been suggested about twelve years before by one Renaud, a young Frenchman who had never seen an action. To increase the effect of the bomb vessels that were sent with the fleet on this occasion, a new galliot of about 300 tons burthen was so contrived as to be itself one great bomb, capable of being discharged wherever she could float. In the hold of this galliot, next the keel,

---

1 This detail is interesting as showing the personnel necessary for the efficient working of 8 mortars, especially as regards fireworkers and bombardiers.

were stowed 100 barrels of powder, and as the effect of powder is always in proportion to the resistance, this layer was covered with a flooring of thick timber, which was perforated in several places to admit the train that was to communicate the fire. Upon the top of this floor was laid 300 carcasses consisting of grenades, cannon bullets, chain-shot, great bars of iron, and an incredible variety of other combustible matter. This galliot was called the "Infernal." After several days bombarding the city, the 19th November preparations were made for playing off the "Infernal." An engineer being put on board, carried her under full sail to the foot of the wall, where she was fired, notwithstanding all the fire of the place against him: but the wind suddenly veering forced him off before the vessel could be secured, and drove her upon a rock within pistol shot of the place where she was to have been moored. The engineer set fire to the train and left her, when she blew up with an explosion that shook the whole city like an earthquake, uncovered about 300 houses, threw down the greater part of the wall towards the sea, and broke all the glass, china, and earthware for three leagues round. The consternation of the people was so great that a small number of troops might have taken possession of the place without resistance; as it was, they demolished Quince Fort, carried off eighty prisoners, and awakened the whole French nation from their golden dreams of the empire of the sea.

1693. 20 Dec. The sum of £210,773 4*s*. 5*d*. allowed for the extraordinary charge of the Office of Ordnance in relation to the land forces for the year 1694.

1694. 7 May. Her majesty's warrant to Henry, Earl of Romney, master-general of the ordnance, directing a train of brass ordnance and several bomb ships be fitted with mortars, &c., and machine vessels, together with a fitting proportion of ammunition and other stores and habiliaments of war, with the officers, fire-masters, fire-workers, bombardiers, gunners, and artificers, forthwith to be provided and fitted out to sea for our service.

Nature of Ordnance.

Culverins	8
Demi-culverins	4
Sakers	6
Falcons	6
10-inch mortars	2
Petards fixed	4

List of officers, gunners, and other attendants appointed to this train, with their pay per diem :—

		£	s.	d.
Colonel	Sir Martin Beckman, Kt.........	2	0	0
Lieut.-colonel	Jacob Richards (paid on the Flanders Establishment)			
Major	John Henry Hopkey	0	15	0
Chaplain..................	William Galloway	0	8	0

For the Train of Artillery.

		£	s.	d.
Commissary	Samuel Pelling	0	10	0
Fire-worker	Henry Brass	0	5	0
3 bombrs. at 2s. 6d. each		0	7	6
Lieutenant	William Bousfield	Detached out of the Flanders Train and paid on that Establishment.		
Gentleman of the ordnance	William Gunn			
3 sergeants of the gunners				
28 gunners				
2 gunners at 2s. each		0	4	0
15 conductors, at 3s. each		2	5	0
Master-carpenter		0	4	0
9 carpenters at 2s. 6d. each		1	2	6
Master-wheelwright		0	4	0
3 wheelwrights, at 2s. 6d. each		0	7	6
Master-smith		0	4	0
2 smiths, at 2s. 6d. each		0	5	0
Collar-maker		0	3	0
Master-tinman		0	4	0
Assistant		0	2	6
Bridge-master	Ambrose Roberts	0	4	0
Assistant	John Wells	0	2	6

For the Bomb Vessels.

		£	s.	d.
Commissary and paymaster	Mathew Purcel	0	10	0
Commissary	David Lewis	0	8	0
Surgeon	Francis Tomkies	0	17	6
His mate	Charles Tomkies			
Four fire-masters at 10s. each	John Fawcett, Thos. Silver, John Baxter, Zachariah Smith	2	0	0
20 fire-workers, at 5s. each	Thos. Heyden Wm. Wilks Andw. Hydeman Thos. Newton Bryan Orach Thos. Dyson John Humphrys Corn. Ambross Wm. Wood, jun. John Winskell Ulrich Zuck Wm. Mesmore Thos. Hall Richd. Silver John Smith Thos. Ball Bennedict Smith Benj. Scarlett Evan Protheroe James Nixon	5	0	0

	£	s.	d.
36 bombardiers, at 2s. 6d. each	4	10	0
Master-carpenter	0	5	6
16 carpenters, at 2s. 6d. each	2	0	0
Master-smith	0	4	0
3 smiths, at 2s. 6d. each	0	7	6
4 conductors, at 3s. 4d. each	0	13	4

For the Machine Vessells.

2 fire-workers (one on the Flanders establishment) ...	0	5	0
6 persons to fire the machines, at 10s. each	3	0	0
Clerk of stores, Alexander Eustace	0	10	0
2 conductors, at 3s. each...	0	6	0
Master-carpenter	0	4	0
2 assistants, at 3s. 6d. each	0	7	0
Tinman and his assistant	0	6	6
Cooper and do.	0	6	6
3 labourers, at 3s. 4d. each	0	10	0

1694. 7 June. Colonel Sir Martin Beckman present with Lord Berkley at the attack of Brest.[1]

St. James's, 21 July, 1694.

21 July. Sir,—I have received yours of the 17th from before Havre de Grace, and am glad to find by it that our bombs have had so good effect, as well upon that town as upon Diep, and that you have escaped the dangers which you have been exposed to upon those occasions.

I have by this post writ to Colonel Richards to come up to town with all possible haste, to give a more particular account to her majesty of those actions; and believing this may find you at St. Helen's, I would have you give the necessary orders for the refitting of the bomb vessels which are become unserviceable.

Whatever mortars we have here that may be of use shall be sent down to you with the first opportunity.

Your faithful servant,

ROMNEY.

Sir Martin Beckman.

19 Sept. The British artillery, under Lieutenant-Colonel Brown, were at the siege of Huy, which surrendered the 19th September. 120 pieces of cannon and mortars were employed at the siege. (Mr. Borgarde was present at this siege.) This siege was particularly remarkable for the fine and numerous artillery which was employed at it, and the dexterity with which it was managed.

---

[1] This train accompanied an expedition of 6000 men sent against Camaret, a small neck of land at the north of the river of Brest, and which would have commanded that river if we could have made ourselves masters of it, but owing to delays in getting to sea the design miscarried, and the expedition returned to Plymouth, where the troops were landed, and the fleet, with the bomb vessels, sailed again to the French coast, where they bombarded Dieppe and Havre de Grace, but failed before Dunkirk.—See *Burnet's History*, Vol. II.

The same number of bombardiers, petardiers, and gunners employed, and paid at the Tower, as for several years previous.

1694. 10 Oct. Her majesty's warrant to Henry, Earl of Romney, master-general of the ordnance, directing a train of brass ordnance and mortars to be in readiness, with all fitting equipage, to be forthwith transported to the West Indias for our service there.

Nature of Ordnance.

24-pounders	4
Culverins	2
Sakers	6
11-inch diameter mortar-pieces ...	2
Brass petards, fixed in frames ...	4

List of officers, gunners, and other attendants appointed to the train of artillery for the West Indias :—

		Per diem. £	s.	d.
2 engineers	Christian Lilly, first engineer & commander-in-chief ...	0	10	0
	Hugh Syms, second engineer ...	0	10	0
Fire-worker, Henry Brass ...		0	5	0
Master-gunner ...		0	5	0
His mate ...		0	3	0
12 gunners, at 2s. each ...		1	4	0
6 bombardiers, at 2s. 6d. each ...		0	15	0
6 miners, at 2s. 6d. each ...		0	15	0
4 carpenters, at 2s. 6d. each ...		0	10	0

19 Dec. Her majesty's warrant for embarking a detachment of artillery for Ostend.

Gentleman of the ordnance, Wm. Gunn.
2 sergeants.
1 corporal.
24 gunners.

1695. From the Quarter Books :—

Sir Martin Beckman...Chief engineer.
Jacob Richards .... do.
Thos. Phillips .... do.
John S. Schlundt ......Chief fire-master
John H. Hopeke ......Mate do.

22 May. Warrant by the Lords Justices: Whereas the king's most excellent majesty hath thought it requisite and necessary for his service that several bomb vessels, with mortars, and a fitting proportion of ammunition and other stores, should be provided for his fleet in the Straights, they authorize the Master-General (Earl of Romney)

to provide and send them to the Straights, with the annexed list of officers, &c.[1]

Brass mortars, 13-inch 12

List of Officers, &c., with their Pay per diem.

		£	s.	d.
Colonel, Sir Martin Beckman, Kt.		2	0	0
Major, John Henry Hopkey		0	15	0
Chaplain, William Galloway		0	8	0
Adjutant, John Hanway		0	8	0
Commissaries,	Mathew Pearcel and paymaster ...	0	10	0
	David Lewis	0	8	0
Surgeon, Francis Tomkiss		0	7	6
His mate, Charles Tomkiss				
Zach. J. Smith ...	3 fire-masters, at 10s. each ...	1	10	0
Thos. Heydon ...				
Geo. Cassell ...				
Wm. Wilks	14 fire-workers, at 5s. each ...	3	10	0
Andw. Hideman ...				
Thos. Dyson				
Ulrich Zuck				
Wm. Mesmore ...				
Thos. Hall				
Richd. Silver ...				
John Winch				
Jas. Nixon				
Alex. Silver				
Chas. Hall				
Joakim Miller ...				
Thos. Dodge				
Hy. Higgins				
36 bombardiers, at 2s. 6d. each		4	10	0
8 conductors, at 3s. each		1	4	0
Master-carpenter		0	6	3
His mate		0	3	0
12 carpenters, at 2s. 6d. each		1	10	0
Master-smith		0	4	0
3 smiths		0	7	6

29 June. A second warrant, directing the fitting out of several bomb and machine vessels for the Channel, with an annexed list of officers, &c.[2]

---

1 These bomb vessels accompanied the fleet under Admiral Russell, which dominated the Mediterranean all the summer, and blockaded Toulon and Marseilles, and caused the French to abandon Palamo in Catalonia.

2 This expedition appears to have bombarded St. Malo and Granville. They also attempted Dunkirk, but failed in the execution. Some bombs were thrown into Calais, but without any great effect, so that the French did not suffer so much by the bombardment as was expected. The country indeed was much alarmed by it; they had many troops dispersed along their coast, so that it put their affairs in great disorder, and we were everywhere masters at sea.—Burnet, Vol. II.

Brass mortars,	13-inch ...	... 18
"	12¼ " ...	... 2
"	10 " ...	... 3

List of Officers, &c., with their Pay per diem.

		£	s.	d.
Colonel Jacob Richards, addition to his Lieutenant-Colonel's pay on the Flanders establishment ...		0	15	0
Adjutant, Michael Richards, paid on the Flanders establishment.				
Pay-master and commissary, Edward Gibbon		0	12	0
5 clerks of stores, at 4s. each		1	0	0
3 fire-masters, at 10s. each	John Fawcet John Baxter Wm. Ward	1	10	0
14 fire-workers, at 5s. each	Thos. Newton John Winskell John Smith Bryant Orack John Humphreys Cornls. Ambrose Wm. Ward Thos. Bell Benj. Scarlett Evan Protheroe Benedict Smith Geo. Murrell Leod. Jackson Wm. Lightfoot	3	10	0
44 bombardiers, at 2s. 6d. each		5	10	0
Master-carpenter		0	5	0
2 mates at, 3s. each		0	6	0
12 carpenters, at 2s. 6d. each		1	10	0
Master-smith		0	4	0
2 smiths, at 2s. 6d. each		0	5	0
Turners		0	4	0

For the Machine Vessels.

Clerk of the stores and pay-master, Alexr. Eustace ...	0	10	0
His assistant, John Snow	0	7	6
3 conductors, at 3s. each	0	9	0
2 labourers, at 3s. 4d. each	0	6	8
8 firemen, at 10s. each	4	0	0
Chief fire-worker, Thomas Barnet	0	10	0
10 fire-workers, at 5s. each	2	10	0
Master-smith	0	4	0
His assistant	0	2	6
Master-cooper	0	4	0
His assistant	0	2	6
Master-tinman, paid on the Flanders establishment.			
His assistant	0	4	0
Master-carpenter	0	5	0
2 mates, at 3s. 6d. each	0	7	0
57 carpenters, at 2s. 6d.	7	2	6

1693. 3 July. The siege of Namur planned and executed by King William in person, one of the strongest places in the Netherlands, and extolled as a masterpiece of the military art, commenced the 3rd July, within sight of the enemy's united forces: it capitulated the 1st September, although defended by a marshal of France, with an army of 15,000 chosen troops for his garrison, and 120 pieces of cannon, with another army more numerous than the allied army in view to relieve it.

The British artillery were extremely active at the siege, commanded by Colonel Goor, second in command Lieutenant-Colonel Brown, who threw a shell into the enemy's magazine on a demi-bastion, and blew the whole up. Mr. Borgarde was present at this siege.

4 July. Colonel Jacob Richards commanded five bomb vessels, under Lord Berkly, at the bombardment of St. Malos: they fired 900 bombs, with great success.

8 July. Granville destroyed by Colonel Richards, with eight bomb vessels, assisted by frigates.

1 Aug. Dunkirk bombarded by Colonel Richards and Lord Berkley's fleet.

9 Aug. Colonel Sir Martin Beckman was employed at the siege and bombardment of Palamos, commanding the artillery with the army under Brigadier-General Stewart, on board Admiral Russell's fleet.

17 Aug. Calais bombarded by Colonel Richards and Lord Berkley's fleet, the English artillery throwing 600 shells from their bomb vessels.

10 Dec. His majesty directs that a company of pioneers, to belong to his train of artillery in Flanders, be raised and continued on constant pay.

Strength:—1 captain, 4 sergeants, 54 privates.

Signed, ROMNEY.

1696. 1 Jan. Mr. Borgarde promoted to captain and adjutant to the train of artillery in Flanders, *vice* Cotten.

1 Feb. Lieutenant-Colonel Holcraft Blood appointed second engineer.

Captain Richard Leake, master-gunner of England, died: he was buried in Woolwich church.

1 April. His majesty's warrant for an augmentation of 4 fire-masters, 20 fire-workers, and 30 bombardiers to the train of artillery.

July. Colonel Sir Martin Beckman, with his bomb vessels, threw 2000 bombs into St. Martins and Olonna.

Four companies of artillery with the army in Flanders.

1696. 28 Oct. Expenses of the ordnance laid before the House for the year 1697.

For the train £210,773 4s. 5d.

30 Oct. George Brown. Esq., appointed, by his majesty's warrant, master-gunner, as well within the Tower of London as within our kingdom of England, with power and emoluments belonging to the said office as fully and in as large and ample manner and form as Anthony Stephen Bull, William Bull, William Hamond, John Reynolds, James Weymes, Captain Vallentine Pyne, and Captain Richard Leake, or any other, &c.

	Capt. Talbott Edwards, engineer, ordered to	Barbadoes,	23 Janry.
1697.	Capt. Heber Lands, „ „	Bermuda,	12 Janry.
	Capt. Henry Brabant, „ „	New York,	9 Febry.

15 Mar. His majesty's warrant to the Earl of Romney, master-general, directing a train of brass and iron ordnance and several brass mortar pieces, with a fitting proportion of ammunition and other stores, to be in readiness for Newfoundland, with the officers, &c., as per list.

Brass ordnance ...	Sakers	6
Iron do.	Demi culverins	12
	Minions ...	6
Brass morar pieces ...	7½-inch ...	3
Brass petards	... ...	6

List of the officers, &c., with their pay per diem :—

			£	s.	d.
Chief engineer and Commander-in-Chief	Michael Richards, besides his pay on the Flanders Train...............	....	0	10	0
Engineer...............	Isaac Francis Petit		0	10	0
3 inferior engineer	Thomas Bell Benj. Withall Francis Hawkins	at 5s. each	0	15	0
1 fire-master	Joachin Miller.......................		0	10	0
2 fire-workers	John Green Thos. Shales	at 5s. each	0	10	0
6 bombardiers, at 2s. 6d. each			0	15	0
1 captain	George Spencer		0	10	0
2 gentlemen of the ordnance	Thos Dodge John Huxford	at 5s. each	0	10	0
3 gunners, at 2s. each			0	6	0
21 practioner gunners, at 1s. 6d. each			1	11	6
Master-carpenter			0	4	0
His mate			0	3	0
10 carpenters, at 2s. 6d. each			1	5	0
Master-smith...			0	4	0
His mate			0	3	0

		£	s.	d.
2 smiths, at 2s. 6d. each		0	5	0
1 surgeon	John Winram	0	5	0
1 mate		0	2	6
1 commissary	David Lewis	0	8	0
1 clerk of stores		0	4	0

1697. July. From Quarter Book. Sir Martin Beckman appointed comptroller of the fireworks, as well for war as tryumph, and of all the fire-masters, fire-workers, bombardiers, and petardiers employed in the Laboratory, at a salary of £200 per annum; John S. Schlundt, chief fire-master, at £150 per annum; John H. Hopkee, mate, at £80 per annum.

27 July. Colonel Woolfgang William Romer ordered to attend Lord Bellamont to his government of New England and New York, as engineer.

19 Oct. Peace proclaimed in London.

The whole of the British artillery in Holland embarked for England.

22 Dec. Pay advanced to the following officers, by order of the Board:—

Lieut.-Col.	George Brown, and as captain	
Chaplain	Wm. Galloway	
Lieutenants	Andrew Bonnell	1st Company
	Mathew Purcell	1st Company
Gentleman of the ordnance	Edwd. Grover	1st Company
Captain	Abraham Cock	2nd Company
Lieutenant	Thos. Baskall	2nd Company
Gentleman of the ordnance	Saml. North	2nd Company
Do. do.	Robt. Braman	2nd Company
Captain	Jonas Watson	3rd Company
Lieut.	Wm. Bousfield	3rd Company
Gentleman of the ordnance	John Hollibone	3rd Company
Captain	Leonard Vanderstam	4th Company
Gentleman of the ordnance	James Finney	4th Company
Fire-master	John Baxter.	
Do.	Robt. Guybon.	
Fire-workers	Francis Cope	
	Wm. Hayes	
	John Speed	
	Percival Dunkin	
	Robt. Palmer	
	John Fletcher	
	Edwd. French	
	John Swiet	
	Alexr. Forbes	
	John Jones	
	Wm. Leathes	
	Abrm. White	
	Thos. Traherne	
	John Kelly	
	Walgrave Beesly	
	Fred Tymme	

31 bombardiers
2 clerks of stores
10 conductors
Capt. Charles Ball
Assistant Francis Ellis
5 wheelwrights
7 smiths
7 collar-makers
7 carpenters
Mustered on the Flanders Train.
Col. Wynant Goor
Lt.-Col. George Brown
Qr.-Master Ralph Wood
Waggon-master Charles Ball
Captain L. Vanderstam
Captain Thos. Morris of Pontoons.

1698. His majesty's warrant, directing two bomb vessels, with mortars, ammunition, and stores, with fire-workers, &c., to sail with a squadron of men-of-war to the Mediterranean, commanded by Mathew Aylmer, Esq.

Mortars 12¾-inch 2

List of fire-workers, &c., with their pay per diem :—

			£	s.	d.
2 fire-workers.........	John Winskell Richd. Silver	at 5s. each	0	10	0
6 bombardiers, at 2s. 6d. each			0	15	0
1 labourer			0	3	4

24 May. Whereas we have thought fit (now that a peace is concluded) to dismiss as well the trayn of artillery that has served us during the late wars in Flanders, as also the several trayns that have been employed in our service by sea. Yet that such persons as have served us well and faithfully during the war, might have some reasonable provision made for their subsistence in time of peace, and for having of a trayn of artillery in greater readiness to march upon any occasion for the necessary defence of our realms and dominions, we have thought fit to direct, that a small trayn of artillery should be composed of such persons as have served us well in the said trayns during the war; and the annexed scheme of such a trayn of artillery having been accordingly prepared, and laid before us for our approbation, we have perused and considered thereof, and do hereby approve of, and establish the same to be entertained in our service and kept in our pay in time of peace untill such time as we shall think fit to signify our further pleasure therein. Our will and pleasure therefore is, and we do hereby authorise and direct you out of such money as shall at any time be paid into the treasury of our ordnance on account of land service, to cause the several sums and yearly allowances mentioned in the said annexed scheme, amounting in the whole to £4,482 10s. to be paid to the respective officers, engineers, gunners, and others therein mentioned, the said allowances to commence from the 1st day of this instant May, and to

continue during our pleasure. And we do hereby further authorise and empower you as often as any vacancies shall happen on this our establishment by the decease of any person now placed thereupon or otherwise, to fill up the same with such persons as have served in any of our above mentioned trayns and could not at present be provided for, or with such other persons as shall apply themselves to study mathematicks, and duly qualify themselves to serve as engineers, fireworkers, bombardiers or gunners on our said establishment, &c.

At our Court at Kensington,
By his Majesty's command,
(Signed) J. VERNON.

To Earl Romney,
Master-General of the Ordnance.[1]

The following was the strength of the regimental train of artillery as established by the foregoing warrant, and the total expense of which was £4,482 10s.

Staff:—

	£	s.	d.
Colonel (on the old establishment)			
Lieut.-Col. (additional to his pay on the old establishment)	55	5	0
Major	50	0	0
Comptroller	45	0	0
Adjutant	60	0	0

1 On the 19th Oct., 1697, the peace of Ryswick was proclaimed in London, and the whole of the British Artillery in Holland embarked for England. Up to this date there was no regular regiment of Artillery, but as we have seen, special trains were equipped by virtue of the royal warrant, and certain officers and ministers were attached to each. The guns of a train were in general worked by the ordinary troops of the line, under the direction of the trained bombardiers, petardiers and gunners, who were drafted from the Tower, or some of the other garrisons of the kingdom. These were permanently maintained on the quarter-books of these garrisons, and were liable to be called upon to serve wherever required, either on board the bomb-vessels of the fleets, or as attached to the trains of artillery. In this manner, we find that there were several distinct trains in the service. When the services of the members of one train were no longer required, the bombardiers and others were drafted for service in other trains. We therefore find that the same officers were generally employed in different services, either in the king's expedition to Ireland to resist James II., or in the Flanders campaign, being transferred from one to the other as circumstances required. In 1693, the several trains in Flanders appear to have been organised in companies, as already mentioned, and on their return to England in 1698, dispersed. Such being the case at the close of the war, the king issued his warrant on the 24th May, 1698, for the establishment of a small regimental train of artillery, composed chiefly of those persons who had served faithfully during the war. On the organisation of this train, it was provided that all such persons who had quitted bombardiers' or gunners' posts in the old Flanders establishment of 1693, to be advanced in the train of 1698, should (in case this latter were reduced) resume their former posts in the old establishment, notwithstanding any warrant granted to other persons for the same.

1st Company:—	£	s.	d.
Captain	100	0	0
First lieutenant	60	0	0
Second lieutenant	40	0	0
2 gent. of the ordnance (paid on the old establishment).			
2 sergeants, 1s. 6d. each	54	15	0
15 gunners on the old establishment.			
15 gunners 1s. each	273	15	0

2nd, 3rd, and 4th Companies the same as the first, with the exception of two gent. of the ordnance who were paid on the new establishment at the rate of £40 each.

Engineers:—	£	s.	d.
6 engineers, £100 each	600	0	0
4 sub-engineers, £50 each	200	0	0
2 firemasters, £100 each	200	0	0
12 fireworkers, £40 each	480	0	0
12 bombardiers, £36 10s. each	438	0	0
Grand total	£4,482	10	0

1 May. A regimental trayn of artillery to be kept in pay in time of peace, established from the 1st May, 1698.

Colonel.
Lieut.-Colonel, George Browne.
Major, John Sigmond Schlundt.
Comptroller, James Pendlebury.
Adjutant, Albrecht Borgard.

1st Company.

Captain, Jonas Watson.

Lieutenants:—
1st, Ralph Wood.
2nd, Joseph Durden.

Gentlemen of the Ordnance:—
Edward Silvester.
Nicholas Whitaker.

Sergeants:—
John Carne.
David Thomas.

15 gunners, old establishment:—
Thomas Lincolne.
William Topp.
Philip Dodwell.
William Penry.
William Head.
True Heart.
Thomas Legg.
John Colson.
Gervis Durrant.
Nicholas Dory.
John Ranton.
Charles Pickup.
Samuel North.
John Smithurst.
Michael Crofts.

15 gunners, new establishment:—
John Bonamour.
Samuel Ball.
John Baxter.
James Haddock.
John Marshall.
Thomas Marwood.
William Wright.
John Milner.
John Hirst.
Thomas Lunsford.
William Morris.
John Dale.
William Christian.
John Charter.
Francis Blyton.

2nd Company.

Captain, Edward Gibbon.

Lieutenants:—
1st, Thomas Raskell.
2nd, Andrew Bonnell.

Gentlemen of the Ordnance:—
Mathew Purcell.
Daniel Coenen.

Sergeants :—
John Ellis.
Amos Walshaw.

15 gunners, old establishment :—
Joseph Hone.
John Bateman.
Thomas Taylor.
George Guy.
George Browne.
William Cadogan.
Peter Guitchett.
Henry Pillar.
James Nixon.
Thomas Hall.
James Little.
John Burton.
Charles Binkley.
Phillip Spooner.
Michael Clarke.

15 gunners, new establishment :—
Edward Stevens.
John Jordan.
John Dobson.
John Jeoffries.
Leonard Jackson.
Thomas Jones.
Thomas Serwen.
John Nash.
Richard Morris.
Robert Leslie.
John Evans.
John Pearne.
Thomas Hughes.
John Bourden.
William Lightfoot.

3rd Company.

Captain, Edmond Williamson.

Lieutenants :—
1st, Peter Gelmuyden.
2nd, Edward Glover.

Gentlemen of the Ordnance :—
Christopher Briscoe.
Thomas Morris.

Sergeants :—
John Collett.
Henry Woodcock.

15 gunners, old establishment :—
John Hollybone.
John Roberts.
Thomas Maynard.
James Murrell.
Hermanus Loobeck.
Thomas Marsh.
Henry Scott.
Samuel Clarke.
Jeremiah Haughton.
Roderick Price.
John Thornton.
William Charter.
Richard Ditchfield.
Henry Moore.
Nathaniel Hill.

15 gunners, new establishment :—
Thomas Wade.
Mathew Pybus.
Thomas Adams.
Edward Norman.
David Nicholson.
John Tonge.
Henry Wooten.
Godfrey Franks.
Moses Jacock.
William Haynes.
William Cahoon.
Charles Allen.
John Hodges.
Barnaby Griffith.
Jonathan Sanderson.

4th Company.

Captain, William Bousfield.

Lieutenants :—
1st, George Brittenstein.
2nd, George Spencer.

Gentlemen of the Ordnance :—
William Gunn
Robert Welch.

Sergeants :—
Bennet Smith.
Edward Rann.

15 gunners, old establishment :—
Fenton Field.
Daniel Warquin.
John Nicholls.
John Thomas.
William Partridge.
Thomas Sealy.
Nathaniel Hobbs.
Benedict Smith.
Thomas Harwood.
John Huxford.
William Browne.

George Fawcett.
Edward Wells.
John Cartwright.
William Wood.

15 gunners, new establishment :—
Cadwallader Mashmore.
Abraham Carpenter.
John Shepheard
Vincent Walker.
George Minnens.
Thomas Suffolke.
David Anderson.
Edward Bacchus.
Thomas Kectian.
Thomas Mitchell.
Alexander Mackbridge.
Charles Blyton.
Nathaniel Norris.
John Winch.
Francis Moseley.

6 engineers (Captains) :—
John Manclere.
Christian Lilly.
Daniel Sherrard.
John Hannaway.
Isaac Francis Petit.
George Conrade Van Cassell.

4 sub-engineers :—
Thomas Bell.
Benjamin Whithall.
Francis Hawkins.
Lucas Boitout.

2 firemasters (Captains) :—
John Lewis Schlundt, chief firemaster,
Robert Guybon. [July '97, at £150.

12 fireworkers :—
Thomas Newton.
Zachariah Smith.
Thomas Heydon.
Andrew Hideman.
William Wilkes.
Bryan Ororch.
Richard Silver.
Thomas Dodge, Jun.
John Winskell.
Alexander Silver.
Joakim Miller.
James Finney.

12 Bombardiers :—
Christopher Wells.
William Oliver.
David Lewis,
Moses Baxter.
James Griffith.
John Speed.
John Jones.
Edward French.
Robert Palmer.
Francis Coape.
Robert Braman.
Alexander Forbes.

1698.					£	s.	d.
27 May.	Lt.-Col. George Brown	appointed	to this train,	with pay per annum	300	0	0
	Major J. S. Schlundt	do.	do.	do.	200	0	0
	Mr. Jas. Pendlebury	do.	Comptroller	do.	100	0	0
	Capt. Alb. Borgard	do.	Adjutant	do.	00	0	0
	Capt. Jonas Watson	do.	1st company				
	Capt. Edwd. Gibbon	do.	2nd do.				
	Capt. Edmond Williamson	do.	3rd do.				
	Capt. Wm. Bousfield	do.	4th do. [1]				

29 Nov. Major J. H. Hopkee appointed to the train under Lieutenant-Colonel Brown, *vice* Major J. S. Schlundt resigned.

---

[1] There is some confusion caused by the double enumeration of the same names, but it appears that the pay of the staff and companies, &c., was obtained under two distinct votes, that for the "old" and for the "new" establishments, in about equal proportions, the whole aggregating £4,482 per annum.

30 Jan. 1698—9. This establishment was broken up on the 31st January, 1699, by his majesty's warrant of the 30th January, and the following persons granted a certain yearly allowance from the 1st February, amounting to £1400, per his majesty's 14 Feb. warrant of the 14th February.

Certain yearly allowances to the officers and others of the regimental train of artillery lately broke:—

		£
4 captains	Jonas Watson Ed. Gibbon Ed. Williamson Wm. Bousfield	at 60 each.
2 fire-masters	John Lewis Schlundt Robt. Guybon	at 60 each.
4 first lieuts.	Ralph Wood Thos. Raskell Peter Gelmoyden Geo. Brittenstein	at 50 each.
5 second lieuts.	Joseph Durden Andrew Bonnell Ed. Glover Geo. Spencer Roger Coleburne	at 40 each.
12 fireworkers	Thos. Newton Zachah. Smith Thos. Heydon Wm. Wilks Bryan Ororch Andrew Hideman Richard Silver Thos. Dodge John Winskell Alex. Silver Joakim Miller Jas. Finney	at 40 each.
8 sergeants	John Carne John Ellis Bennet Smith Ed. Rann Henry Woodcock John Collett David Thomas Amos Walshaw	at 20 each.

14 Feb. His majesty by warrant ordered an additional establishment to be kept in pay in the time of peace, to commence the 1st February.

	£	s.	d.
6 engiueers, at £100 each..	600	0	0
4 sub-engineers, at £50 each	200	0	0
6 gentlemen of the ordnance, at £40 each	240	0	0
12 bombardiers at £36 10s. each	438	0	0
60 gunners, at £18 5s. each	1090	0	0
Total............. ...	£2073	0	0

1698—9. 27 Mar. The following engineers were appointed to this establishment:—

Capt. John Mancleere, Albert Borgard, Dan. Sherrard, John Hannaway, Isaac Fran. Petit, George Conrade — as engineers.

Thos. Bell, Benj. Withall, Francis Hawkins, Lucas Boitout — sub-engineers.

Mathew Purcell, Danl. Coenen, Christ. Briscoe, Thos. Morris, Wm. Gunn, Robt. Welsh — gentlemen of the ordnance.

1700 Jan. Quarter Book:—

	Per annum. £	s.	d.
Sir Martin Beckman, chief engineer, salary	300	0	0
Col. Holcraft Blood 2nd do.	250	0	0
Col. Jacob Richards 3rd do.	150	0	0
Thos. Phillips 4th do.	100	0	0
George Brown, master-gunner of England	190	0	0
John Leake, Thos. Dodge, Thos. Silver — Mates to do. ... ... each	45	10	0
Sir Martin Beckman, comptroller of the fireworks	200	0	0
Major Henry Hopkee, chief fire-master...	150	0	0
Captain John Baxter, mate to do.	80	0	0
Captains James English, Jacob Blayvitz, Willm. Wood, and James Pendlebury, fireworkers and petardiers — each	40	0	0
Captain Charles Ball, waggon-master	100	0	0
John Blake and John Allen, gentlemen, proof-masters, each ...	20	0	0

1700. The bombardiers, petardiers, and gunners paid for working at the Tower, the same in number as for many years past.

"To Talbott Edwards and Peter Carles, engineers, for their encouragement to travel into foreign parts to perfect themselves in the arts of fortification and other parts of the mathematicks as may render them capable of serving his majesty, each £100 per annum."

The Board ordered the following to be reimbursed for taxes assessed on their salaries for the year 1699:—

4 first lieutenants.
5 second lieutenants.
12 fireworkers.
4 sub-engineers.
6 gentlemen of the ordnance.
8 sergeants.
23 bombardiers.
4 petardiers.
122 gunners.
4 armourers.
64 labourers.

1700. 15 April. His majesty's warrant directing the Carcass bomb vessel to be fitted with two mortars, ammunition, and stores, with a fireworker and bombardiers, to be sent with the squadron under Sir George Rooke.

Mortars, 13-inch.........2

		£	s.	d.
Fireworker	Alex. Silver	0	5	0
4 bombardiers at 2s. 6d. each		0	10	0

1702. 5 Feb. An English army with artillery sent to Holland.

Estimate for the Ordnance land service for the year:—

		£	s.	d.
Holland Train.	Ammunition	13,000	0	0
	Pay of Train officers	9,000	0	0
	Horses and wagons	12,000	0	0
	Contingencies	1,000	0	0
	Ordinary of the office	28,273	13	9
	Saltpetre	7,700	0	0
		£70,973	13	9

14 March. Her majesty's warrant directing a train of artillery with ammunition and attendants to be transported to Holland.

		No.
Brass ordnance.	Sakers	14
	3-pounders	16
	Howitzers	4

List of officers, &c., with their pay per diem :—

	£	s.	d.
Colonel, George Brown, appointed 1st April	1	5	0
Lieut.-colonel	1	0	0
Comptroller	1	0	0
Major, Jonas Watson, appointed 1st April	0	15	0
Paymaster	0	10	0
His assistant	0	3	6
Adjutant	0	6	0
Quartermaster	0	6	0
Chaplain	0	6	0
Commissary of the horse...	0	6	0
Master surgeon	0	5	0
His assistant	0	3	0
Provost marshall	0	2	6
Kettle drummer	0	3	0
Kettle driver	0	1	6

First Company of Gunners.

	£	s.	d.
Captain	0	10	0
Lieutenant...	0	6	0
Gentleman of the ordnance	0	4	0
3 sergeants at 2s. 6d. each	0	7	6
3 corporals at 2s. each	0	6	0
25 gunners at 1s. 6d. each	1	17	6
25 mattrosses at 1s. each	1	5	0

Second Company of Gunners.

	£	s.	d.
Captain	0	10	0
Lieutenant...	0	6	0
Gentleman of the ordnance	0	4	0
3 sergeants at 2s. 6d. each	0	7	6
3 corporals at 2s. each	0	6	0
25 gunners at 1s. 6d. each	1	17	6
25 mattrosses at 1s. each	1	5	0
2 fireworkers at 4s. each	0	8	0
8 bombardiers at 2s. each...	0	16	0
Commissary of stores	0	6	0
Assistant	0	4	0
2 clerks at 4s. each	0	8	0
11 conductors at 2s. 6d. each	1	7	6
1 conductor and cooper	0	3	0
Master carpenter	0	4	0
His mate	0	3	0
6 carpenters at 2s. 6d. each	0	15	0
Master wheelwright	0	4	0
3 wheelers	0	7	6
Master smith	0	4	0
2 smiths at 2s. 6d. each	0	5	0

	£	s.	d.
Master tinman	0	4	0
1 tinman	0	2	6

A Company of Pioneers.

	£	s.	d.
2 sergeants at 2s. 6d. each	0	5	0
20 pioneers at 1s. each	1	0	0

A Company of Pontoon Men.

	£	s.	d.
Bridge-master	0	5	0
2 corporals at 2s. each	0	4	0
20 pontoon men at 1s. 6d. each	1	10	0

1702. 6 April. Her majesty's warrant to the Earl of Romney, master-general, directing him to fit out a train of artillery to attend our land forces, ordered to embark on an expedition commanded by James, Duke of Ormond; also several bomb vessels to attend the fleet for this summer's service. (They attacked the French fleet and the Spanish galleons in Vigo.)

Brass ordnance.	24-pounders	10
	18 do.	10
	$5\frac{1}{4}$ do.	4
	3 do.	6
	13-inch mortars	6
	10-inch do.	6
	$7\frac{1}{4}$-inch do.	4
	Hand mortars	10
Brass	Petards	10

List of officers, &c., with the train, and their pay per diem:

			£	s.	d.
Colonel	George Brown	apptd. 10th April	1	5	0
Major	Albert Borgard		0	15	0
Adjutant	Thomas Hall		0	6	0
16 engineers at 10s. each			8	0	0
Paymaster	John Pearne		0	8	0
Commissary of stores	Alex. Eustace		0	8	0
His clerk	Henry Blake		0	4	0
9 conductors of stores			1	2	6
Surgeon	— Hardwick		0	5	0
His mate	John Beale		0	3	0
4 pontoon men at 1s. 6d. each			0	6	0

A Company of Gunners.

		£	s.	d.
Captain	Robt. Guybon	0	10	0
First lieut.	Richd. Silver	0	6	0
Second lieut.	Walgrave Beezly	0	5	0

			£	s.	d.
2 gentlemen of the ordnance	Thos. Traherne, Alex. Hara	at 4s. each...	0	8	0
4 sergeants at 2s. 6d. each			0	10	0
4 corporals at 2s. each			0	8	0
30 gunners at 1s. 6d. each			2	5	0
60 mattrosses at 1s. each			3	0	0
8 fireworkers	Mathew Pybus, Wm. Haynes, Thos. Mitchell, Vincent Walker, John Cartwright, Henry Pillir, Thos. Bassett	at 4s. each	1	12	0
16 bombardiers at 2s. each			1	12	0
Master carpenter			0	4	0
6 carpenters at 2s. 6d. each			0	15	0
Master wheelwright			0	4	0
4 wheelwrights at 2s. 6d.			0	10	0
A cooper			0	3	0
2 farriers or smiths at 2s. 6d. each			0	5	0
2 collar-makers at 2s. 6d. each			0	5	0
Master tinman			0	3	0
1 tinman			0	2	6

A Company of Miners.

	£	s.	d.
1 lieutenant	0	4	0
20 miners at 1s. 6d. each	1	10	0
Commissary of horse and paymaster's assistant	0	6	0
Assistant to the commissary	0	3	0
3 conductors of horses at 2s. 6d. each	0	7	6
25 carters at 1s. 6d. each	1	17	6

Ordnance and Officers, &c., for the Bomb Vessels.

Brass Ordnance ... 13-inch mortars 7

Colonel	George Brown
Major	Albert Borgard
Fire-masters	Lewis Schlundt, Thos. Heydon
16 fireworkers	William Wilks, Andrew Hydeman, Thos. Dyson, Percival Donkin, John Green, Nicholas Devey, Leonard Jackson, Evan Protheroe, William Wood

Thos. Hughes
Cadwalader Masmoir
John Smith
Benj. Scarlett
John Winch
Wm. Lightfoot
Henry Wotton

32 bombardiers
Commissary........................... George Hay
Assistant .... Chas. Buckley
9 conductors
Master carpenter
His mate
15 carpenters

Captain Borgard promoted to major; he embarked and sailed with five bomb vessels in the expedition to Cadiz and Vigo.

1702. 10 April. Armour delivered for the service of the train now going to sea.

Harquebuss armour, carbine proof: Backs 10, breasts 10, potts 10.

12 May. The Earl of Marlborough appointed master-general. He embarked to command the allied army in Holland.

31 July. The British artillery joined the allied army in camp at Geldorp.

29 Aug. The British army and artillery reviewed by the Earl of Marlborough at Breda.

5 Sept. Colonel Blood present at the taking of Venlo, where he acted as first engineer under General Cohorn.

23 Oct. Also at Liege.

30 Dec. Her majesty's warrant to the Duke of Marlborough, M.G., to cause two brass mortars and a sufficient quantity of stores to be provided and transported to the West Indias for our service there, under the command of Brigadier Colenbine; likewise two engineers at ten shillings each, and three bombardiers at two shillings each per diem, to attend the said expedition.

Brass mortars, 7¼-inches 2

1702. 8 Feb. Her majesty's warrant to augment the train of artillery sent to Holland on the 14th March last, with brass ordnance, mortars, ammunition, and other stores, also officers, gunners, bombardiers.

Nature and Number of Ordnance.

Brass	Howitzers	4
	Demi-culverins	6
	Sakers	14
	3-pounders	16

List of officers, &c., with their pay per diem :—

			£	s.	d.
Master-general					
His secretary					
His clerk					
Colonel	Holcraft Blood	Appointed 20th February	1	5	0
Lieut.-colonel	John Hopke	Appointed 20th February	1	0	0
Comptroller & Lieut.-Colonel	James Pendlebury	Appointed 24th February.	1	0	0
Major	Jonas Watson, 20th February		0	15	0
Paymaster			0	10	0
His assistant			0	3	6
Adjutant			0	6	0
Quartermaster			0	6	0
Chaplain			0	6	0
Commissary of horse			0	6	0
Master surgeon			0	5	0
His assistant			0	3	0
Provost marshall			0	3	6
Kettle drummer			0	3	0
Kettle driver			0	1	6

Two Companies of Gunners consisting of, each,

1st Company.

	£	s.	d.
1 captain	0	10	0
Lieutenant	0	6	0
Gentleman of the ordnance	0	4	0
3 sergeants at 2s. 6d. each	0	7	6
3 corporals at 2s. each	0	6	0
30 gunners at 1s. 6d. each	2	5	0
25 mattrosses at 1s. each	1	5	0
Second Company (as above)	5	3	6
2 fireworkers at 4s. each	0	8	0
8 bombardiers at 2s. each	0	16	0
Commissary of stores	0	6	0
His assistant	0	4	0
2 clerks at 4s. each	0	8	0
11 conductors at 2s. 6d. each	1	7	6
1 conductor and cooper	0	3	0

	£	s.	d.
Master carpenter	0	4	0
His mate	0	3	0
6 carpenters at 2s. 6d. each	0	15	0
Master wheelwright	0	4	0
3 wheelwright at 2s. 6d. each	0	7	6
Master smith	0	4	0
2 smiths at 2s. 6d. each	0	5	0
Master tinman	0	4	0
1 tinman	0	2	6

A Company of Pioneers consisting of

2 sergeants at 2s. each	0	4	0
20 pioneers at 1s. each	1	0	0

A Company of Pontoon Men consisting of

A bridge-master	0	5	0
2 corporals at 2s. each	0	4	0
20 pontoon men at 1s. 6d. each	1	10	0

Making four companies of artillery in Holland.

1702. 30 March. Two warrants from her majesty, one for the fitting two bomb vessels with mortars and ammunition for the channel service; the second, several bomb vessels with mortars and ammunition to attend the fleet ordered to the Mediterranean under the command of Sir Cloudsley Shovell.

Brass mortars	For the Channel service	13-inch	3
	For the Mediterranean	13-inch	7

Brass mortars, with stores for the land forces going with the squadron to the Mediterranean	13-inch	2
	10-inch	2

Fireworkers, &c., for the Channel Service.

	Per diem. £	s.	d.
3 fireworkers at 4s. each	0	12	0
6 bombardiers at 2s. each...	0	12	0
2 carpenters at 2s. 6d. each	0	5	0

Officers, &c., for four Bombs for the Mediterranean.

Fire-master	0	8	0
7 fireworkers at 4s. each	1	8	0
14 bombardiers at 2s. each	1	8	0
Commissary of stores	0	6	0
4 conductors of stores at 2s. 6d. each	0	10	0
Master carpenter	0	4	0
4 carpenters at 2s. 6d. each	0	10	0

Persons appointed to attend the Mortars for Land Service.

	£	s.	d.
Fire-master	0	8	0
4 fireworkers at 4s. each	0	16	0
8 bombardiers at 2s. each	0	16	0

1702. April. Thomas Silver, Esq., appointed master-gunner of England.

ANNE R.

30 June. Right trusty and right well beloved cousin and councillor, we greet you well. Whereas King Charles the Second, and King James the Second, did make an establishment of the office of Ordnance, in a book intituled, Instructions for the Government of the Office of our Ordnance under our Master-General thereof, committed to five principal officers, and signed by the said King Charles and King James, and the principal secretaries of state, which were confirmed by his late majesty King William; our will and pleasure, and express command is, That all orders, warrants, and instructions, or directions therein contained and given by the said King Charles and King James as aforesaid, shall continue in force to be obeyed and observed by all and every person or persons who in our said office are or shall be concerned, as fully and amply as the same was or ought to be observed in the reigns of the said King Charles, King James, and King William, and for your so doing this shall be your sufficient warrant, untill our further pleasure be known. Given at our Court at St. James's, this 30th day of June, 1702, in the first year of our reign.

By Her Majesty's command,

NOTTINGHAM.

To John, Earl of Marlborough, our Master-General of our Ordnance.

1703. Sir Jonas Moore, late master-surveyor of the Ordnance, published many works on modern fortification, military architecture, and defending and besieging forts, &c., in this year.

Estimate for Ordnance land service :—

	£	s.	d.
Train in Holland, &c.	35,000	0	0
Ordinary of the office.........................	28,273	13	9
Saltpetre	7,700	0	0
	£70,973	13	9

1704. Estimate Ordnance land service:—

	£	s.	d.
Holland Train	35,000	0	0
Ordinary of the office	28,273	13	9
Saltpetre	10,600	0	0
Repairing of fortifications	15,000	0	0
Extra charge of bedding	1,000	0	0
Expedition to Portugal	28,488	19	9
	£118,362	13	6

1705.
1706.
1707. Estimate Ordnance land service:—

	£	s.	d.
For all provisions ... each	120,000	0	0

1708. Estimate Ordnance land service:—

		£	s.	d.
For all provisions		120,000	0	0
Spanish Train	Stores	18,184	2	3
	Pay of officers	10,411	12	6
	Pay of pioneers	808	10	0
	Contingencies	1,000	0	0
		£150,404	4	9

1703. 13 June. Her majesty's warrant to fit out a bomb vessel with two mortars and ammunition for Channel service.

Brass mortars, 13 inch. 2

	Per diem.		
	£	s.	d.
2 fireworkers at 4s. each	0	8	0
4 bombardiers at 2s. each	0	8	0

24 July. Her majesty's warrant for fitting out a train of artillery with its ammunition and stores for Portugal.

Brass ordnance, sakers 5

List of officers, &c., with their pay per diem:—

	£	s.	d.
Commander in chief, Albert Borgard, apptd. 7th Aug.	1	0	0
6 engineers at 10s. each	3	0	0
Commissary of stores	0	6	0
5 bombardiers at 2s. each	0	10	0
20 gunners at 1s. 6d. each	1	10	0
10 miners at 1s. 6d. each	0	15	0
Master smith	0	4	0
His mate	0	3	0
Master wheelwright	0	4	0
His mate	0	3	0

	£	s.	d.
Master cooper	0	4	0
Master locksmith	0	4	0
His mate	0	3	6
Master carpenter	0	4	0
His mate	0	3	0

1703. 2 Aug. Her majesty's warrant for fitting out two brass guns (9-pounders), with a suitable proportion of ammunition and stores for our service beyond seas, and a proper person for their care at an allowance of 2s. 6d. per diem.

Another warrant for fitting out two bomb vessels with mortars and ammunition for Portugal or elsewhere, and a proportion of fireworkers and bombardiers, &c.

Major Borgarde, appointed to command the artillery, embarked for Portugal. One company of artillery on this service.

The British artillery took the field with the allied army in the Netherlands.

24 April. The allied army besiege Bonn, their artillery consisting of 100 large cannon, 36 mortars, besides 500 Cohorn mortars; it surrendered the 15th May.

Captain Richards, commanding at St. Johns, Newfoundland, joined the fleet for an attack on Placentia, where he acted as engineer.

1704. 7 Jan. Her majesty's warrant for supplying the trains of artillery in Holland with stores.

16 Jan. Her majesty's warrant directing an augmentation to the train in Holland, also ammunition and other stores of war, with officers, &c., to be sent over for our service.

Brass ordnance	
Culverins	6
3-pounders	4

	Per diem. £	s.	d.
2 gentlemen of the ordnance at 4s. each	0	8	0
16 gunners at 1s. 6d. each	1	4	0
60 mattrosses at 1s. each	3	0	0
2 wheelwrights at 2s. 6d. each	0	5	0

18 May. The British artillery joined the allied army in camp at Bedburgh.

2 July. They fought the battle of Schellenberg, near Donawert.

1704. 24 July. Gibraltar taken by the fleet under Sir G. Rooke.

13 Aug. The British artillery in the battle of Blenheim; the confederates had 52 pieces of cannon in the field. Col. Blood distinguished himself by keeping the enemy's foot in check with his nine field pieces loaded with cartridge shot. Upwards of 100 cannon were taken from the enemy.

3 Oct. Her majesty's warrant directing mortars, shells, carriages, ammunition, and other stores be provided and transported to Gibraltar with an engineer, fireworkers, &c.

List of officers, &c., with their pay per diem:—

Brass mortar pieces, 13 inch. 6

	£	s.	d.
Chief engineer	1	0	0
Storekeeper	0	8	0
His clerk	0	4	0
2 fireworkers at 4s. each	0	8	0
6 bombardiers at 2s. each...	0	12	0
55 gunners at 1s. 6d. each	4	2	6

3 Oct. The French and Spaniards besiege Gibraltar, which was defended by Prince D'Armstadt, who received reinforcements of artillery and troops from England.

Major Borgarde promoted to lieutenant-colonel.

1705. 15 Feb. Her majesty's warrant directing several bomb vessels with mortars, ammunition, and stores, officers, &c., to be fitted out for sea to attend our fleet for this year's service, under Sir Cloudsley Shovell.

Brass mortars, 13 inch. 7

	Per diem.		
	£	s.	d.
8 fireworkers at 4s. each	1	12	0
16 bombardiers at 2s. each	1	12	0
4 carpenters at 2s. 6d. each	0	10	0

2 May. Warrant constituting Thomas Erle, Esq., lieutenant-general of all manner of ordnance, munitions, provisions, stores, and habiliments of war within the kingdom of England and town of Berwick-upon-Tweed; and also keeper of the messuage called the Storehouse, belonging to the Ordnance; and also keeper of the Artillery Garden; signed by the Lord Treasurer Godolphin.—*Harl. M.* 2262, 54.

After a siege of six months, the French and Spaniards retired from before Gibraltar on the 22nd March, having expended above eight thousand shells and seventy thousand cannon shot.

1705. Colonel Blood promoted to brigadier-general.

2 May. At the siege and capture of Valencia D'Alcantra, Lieutenant-Colonel Borgarde was wounded, his left arm shot to pieces.

Two companies of artillery in Spain.

12 April. Her majesty's warrant directing a train of artillery to be provided to attend the land forces under the command of the Earl of Peterborough. They sailed from Portsmouth the 24th May. The fleet commanded by Sir Cloudesley Shovell.

Brass	Mortars	2
	3-pounders	8

with a large proportion of stores, ammunition, and carriages for 24-pounders.

List of officers, &c., with their pay per diem:—

	£	s.	d.
Colonel	1	5	0
Adjutant	0	6	0
2 engineers at 10s. each	1	0	0
Commissary and paymaster	0	8	0
4 conductors at 2s. 6d. each	0	10	0
Master gunner	0	5	0
4 sergeants at 2s. 6d. each	0	10	0
4 corporals at 2s. each	0	8	0
10 gunners at 1s. 6d. each	0	15	0
Fire-master	0	8	0
Fireworker	0	4	0
2 bombardiers at 2s. each	0	4	0
Master carpenter	0	4	0
His mate	0	3	0
Master wheelwright	0	4	0
His 2 mates at 3s. each	0	6	0
2 smiths at 2s. 6d. each	0	5	0
Collar-maker	0	3	0

18 April. Lieutenant-Colonel Borgard appointed commander-in-chief of the train of artillery "for Portugal or elsewhere."

3 June. The British artillery took the field and assembled with the allies near Sirk.

3 July. Her majesty's warrant directing gun carriages, powder, shot, and stores to be transported to Newfoundland with the following:—

Demi-culverins	8
Minions	6

Ordnance carriages	Demi-cannon	4
	24-pounders	8
	Culverins	4

	Per diem. £	s.	d.
Gunner and smith...	0	3	0
" " amourer	0	3	0
2 gunners at 1s. 6d. each....	0	3	0
Carpenter ...	0	4	0
Mason	0	4	0

1705. 18 July. In the successful attack of the French lines, by the Duke of Marlborough, on the Maise, eight pieces of cannon with treble barrels[1] were taken from the enemy. Colonel Richards, one of the Duke's adjutants, and who had the direction of making the bridges, and who had behaved himself very well in this action, was sent to the Emperor at Vienna with the welcome news.

10 Aug. Her majesty's warrant directing an additional number of officers, &c., to be sent to Portugal.

Brass ordnance, 3-pounders 5

List of officers, &c., with their pay per diem:—

	£	s.	d.
Captain	0	10	0
Lieutenant...	0	6	0
Gentleman of the ordnance	0	4	0
Fireworker...	0	4	0
Surgeon	0	5	0
Collar-maker	0	3	0
2 sergeants of mattrosses at 2s. 6d. each	0	5	0
Sergeant of miners	0	2	6
42 mattrosses at 1s. each...	2	2	0

11 Aug. The Earl of Peterborough arrives and invests Barcelona.

2 May. Alcantara besieged, also Albuquerque; both surrendered in a few days.

23 Sept. Barcelona taken by the Earl of Peterborough.

Brigadier-General Holcraft Blood, second engineer.
Captain Christian Lilly, third engineer.

---

1 Specimens of these triple cannon (taken at Malplaquet) are now at the R. M. Repository, Woolwich.

1706. 30 March. Her majesty's warrant directing two bomb vessels, with mortars, ammunition, and stores, should be fitted out for sea, with fireworkers, &c., with their pay per diem.

Brass mortars, 13 inch .. ..3.

	£	s.	d.
3 fireworkers at 4s. each	0	12	0
6 bombardiers at 2s. each...	0	12	0
2 carpenters at 2s. 6d. each	0	5	0

10 April. Lieutenant-Colonel Borgarde promoted to colonel. (He was with the artillery at the siege and taking of Alcantara, where he received a contused wound in his left breast; the artillery were also present at the siege and capitulation of Ciudad Rodrigo.)

8 May. The British artillery marched from Breda, and with the English army joined those of the States on the 20th at Grosz Waren.

23 May. They fought the battle of Ramillies, in which the French lost the whole of their cannon; amongst them were twelve pieces of treble cannon.

The confederate artillery consisted of 100 cannon, 20 howitzers, and 42 pontoons.

Colonel Richards, aide-de-camp to the Duke, was sent to England with the news.

1706. 13 May. Her majesty's warrant directing a train of artillery should be embarked to attend the land forces on an expedition commanded by the Earl of Rivers. (They sailed in August and appeared off Lisbon in October—afterwards landed the troops at Alicant.)

Brass ordnance	24-pounders	20
	Culverins	6
	12-pounders	4
	Demi-culverins	4
	Sakers	6

List of officers, &c., with their pay per diem:—

	£	s.	d.
Colonel	1	5	0
4 engineers at 10s. each	2	0	0
2 sub-engineers at 5s. each	0	10	0
Paymaster	0	8	0
Surgeon	0	5	0
His mate	0	3	0
Captain	0	10	0
Lieutenant	0	6	0
2 gentlemen of the ordnance at 4s. each	0	8	0
3 sergeants at 2s. 6d. each	0	7	6
3 corporals at 2s. each	0	6	0

	£	s.	d.
32 gunners at 1s. 6d. each	2	8	0
64 mattrosses at 1s. each	3	4	0
Lieutenant of the miners	0	4	0
17 miners at 1s. 6d. each	1	5	6
Fire-master	0	8	0
3 fireworkers at 4s. each	0	12	0
12 bombardiers at 2s. each	1	4	0
Commissary	0	8	0
Clerk of stores	0	4	0
9 conductors at 2s. 6d. each	1	2	6
3 smiths at 3s. each	0	9	0
Master wheelwright	0	4	0
2 wheelwrights at 3s. each	0	6	0
Master carpenter	0	4	0
4 carpenters at 3s. each	0	12	0
Master cooper	0	4	0
His mate	0	3	0
Collar-maker	0	3	0
2 farriers at 3s. each	0	6	0
3 conductors of horses at 2s. 6d. each	0	7	6
15 carters at 1s. 6d. each	1	2	6

1706. 1 June. Her majesty's warrant directing two bomb vessels with mortars and ammunition to be fitted for sea.

Brass mortars, 13 inch 4

	£	s.	d.
4 fireworkers at 4s. each	0	16	0
8 bombardiers at 2s. each	0	16	0
2 carpenters at 2s. 6d. each	0	5	0

1706. 16 July. Her majesty's warrant. WHEREAS in the Book of General Instructions for the Better Governing of the Office of Ordnance the lieutenant-general and the rest of the principal officers of our Ordnance (or the major part of them) are impowered to transact all matters relating to our said office, in the absence of our master-general, but no payments of money can be made without the lieutenant-general's hand to the said payments, and whereas our affairs at this time call for the attendance of our lieutenant-general of our Ordnance abroad in our army, and that our service in the said office may not suffer or be impeded in any respect by such the absence of our said lieutenant-general, our will and pleasure is, and we do hereby authorise and impower you, the rest of our said principal officers of our Ordnance, (or any three of you), to act and perform all matters in our said office, and to sign for all payments of money there, during the absence only of our said lieutenant-general, as effectually to all intents and purposes whatsoever as if our said lieutenant-general was actually present, and for your so doing this shall be as well to you as to the auditors of imprests, and all other our officers concerned, sufficient warrant.

By Her Majesty's command,

C. HEDGES.

To the Principal Officers of our Ordnance.

1706. April. Lieutenant-Colonel Henry Hopke appointed comptroller of fireworks.

Lieutenant-Colonel James Pendlebury, chief firemaster.

1707. 9 Jan. Supply granted for

	£
Land service of the ordnance ...	120,000
Towards making a wharf and storehouse at Portsmouth...	10,000

1707. 18 Jan. Her majesty's warrant directing the several trains of artillery under the Earls of Galloway, Peterborough and Rivers be reduced into a field train consisting of such brass ordnance, mortars, ammunition, and other stores of war, with such officers, engineers, &c.

		Pay per diem.		
		£	s.	d.
Colonel ...		1	5	0
Lieut.-colonel	... John Hopke, apptd. 1st February	1	5	0
Major ...	Lieut.-Col. Jonas Watson, ,,	0	15	0
Comptroller		1	0	0
Paymaster ...		0	8	0
His assistant		0	3	6
Adjutant ...		0	6	0
Quartermaster		0	6	0
Commissary of the horses		0	6	0
Waggon-master ...		0	6	0
His assistant		0	3	0
Surgeon ...		0	5	0
2 assistants at 3s. each		0	6	0
Provost marshall ...		0	3	0
2 assistants at 2s. each		0	4	0
2 captains at 10s. each		1	0	0
2 lieutenants at 6s. each		0	12	0
2 gentlemen of the ordnance at 4s. each		0	8	0
6 sergeants at 2s. 6d. each		0	15	0
6 corporals at 2s. each		0	12	0
40 gunners 1s. 6d. each		3	0	0
4 drummers at 1s. each		0	4	0
80 mattrosses at 1s. each...		4	0	0
10 engineers at 10s. each...		5	0	0
Fireworker		0	4	0
2 bombardiers at 2s. each...		0	4	0

	Pay per diem.		
	£	s.	d.
Commissary of stores	0	8	0
His clerk	0	4	0
12 conductors of stores at 2s. 6d. each	1	10	0
Master wheelwright	0	4	0
4 mates at 3s. each	0	12	0
Master smith	0	4	0
3 mates at 3s. each	0	9	0
Master gunsmith	0	4	0
2 mates at 3s. each	0	6	0
Farrier	0	3	0
Master cooper	0	4	0
2 mates at 3s. each	0	6	0
Collar-maker	0	3	0
Master carpenter	0	4	0
3 mates at 3s. each	0	9	0

Making two companies, as appears by the paymaster's books.

1707. 25 April. Colonel Borgarde with the artillery were in the battle of Almanza.

22 Dec. The House voted the following supplies:—

	£
Land service of the ordnance	120,000
Storehouse and wharf at Plymouth..............	10,000
Fortifications of Gibraltar	12,284

1708. 31 Jan. Michael Richards appointed colonel of the train of artillery in Spain (under the Earl of Rivers) from 7th of October, 1707.

Albert Borgard, lieutenant-colonel to the train of artillery in Spain.

At £1 15s. per diem each.

Thomas Erle, major to the train of artillery in Spain.

At 15s. per diem.

1 March. Her majesty's warrant directing 4 bomb vessels with mortars and ammunition to be fitted for Channel service.

Brass mortars, 13 inch 6

	Per diem.		
	£	s.	d.
6 fireworkers at 4s. each	1	4	0
12 bombardiers at 2s. each	1	4	0
4 carpenters at 2s. 6d. each	0	10	0

1708. 25 March. Colonel J. H. Hopke appointed first colonel of the artillery in Holland.

James Pendlebury appointed second colonel and comptroller in Holland.

Jonas Watson appointed first lieutenant-colonel in Holland.

Alexander Hara appointed second lieutenant-colonel in Holland.

William Bousfield appointed major in Holland.

7 May. The British forces, with its artillery consisting of 45 pieces, were reviewed at Brussels by the Duke of Marlborough.

Nov. Colonel Michael Richards commanding the artillery in Spain; Captain Bradbury commanding a company.

Lieutenant	D. Warquin.
Gentleman of ordnance	F. Hull.
Quartermaster	John Price.

20 Nov. Her majesty's warrant directing gun carriages, powder, shot, and other stores, with 60 Cohorn mortars to be provided, and ready to be sent by sea upon an expedition for our service, with the following officers, &c.

	Pay per diem.		
	£	s.	d.
1 engineer ...	0	10	0
A commissary	0	10	0
His assistant	0	4	0
A labourer ...	0	3	0
Carpenter ...	0	4	0
Wheelwright	0	4	0

26 Nov. The allied army assembled under the Duke and marched to Bellingham, its artillery consisting of 108 cannon, 24 mortars, and 44 pontoons.

11 July. The Battle of Oudenarde fought. In this battle the Electoral Prince of Hanover (afterwards George the 2nd) disobeyed the Duke's orders in charging the enemy at the head of the Hanoverian cavalry, after being positively refused his request of heading the regiment of Greys in the charge, for which the Duke ordered him to be tried by a court-martial. The Prince appeared, and like Epaminondas of old, he generously and openly confessed his having charged at the head of his father's cavalry. The defence was sustained; he was degraded from the rank of a major-general to that of a corporal, and the report being made to the Elector his father, the same was approved by him. However, the suspension was soon taken

off, the Prince was honorably restored to his dignity, and the army in general, from this sudden eclipse in his station, began to look upon his aspect as more brilliant after the clouds were blown over that had darkened it for a time; and with his employment he regained the favour and affection of his father, of which he seemed in some measure to have been deprived.—*Henderson's History of the Rebellion.*

1708. 22 Aug. Lisle besieged by the allies; their artillery consisted of 120 battering cannon and 40 mortars; it was taken the 23rd October.

25 Aug. Colonel Borgarde commanded the artillery in the attack on Minorca and siege of the castles of Port Mahon, &c., which surrendered the 19th September. In the castles and forts were found 100 pieces of cannon, 3000 barrels of powder, and all things necessary for a good defence. The garrison consisted of 1000 men, half French and half Spanish.

16 Dec. Her majesty's warrant directing a train of artillery "should be forthwith regulated and established at our Castle of Edinburgh."

Nature of Ordnance.

Iron	18 pounders	11
	12 do.	6
	9 do.	7
	6 do.	6
Brass mortars	13 inch	4
	10 do.	2
	7½ do.	1

Officers and gunners, &c., with their pay per diem:—

	£	s.	d.
Captain of gunners	0	8	0
1 engineer	0	7	0
1 lieutenant	0	5	0
1 commissary	0	5	0
1 corporal gunner	0	1	3
10 gunners at 1s. each	0	10	0
6 pontooners at 6d. each	0	3	0
6 bombardiers at 2s. each	0	12	0
1 petardier	0	2	0
2 miners at 1s. 6d. each	0	3	0

At Edinburgh Castle.

1 storekeeper	0	6	1
1 deputy to do.	0	1	8
1 gunsmith	0	2	2

At Stirling.

	£	s.	d.
1 storekeeper	0	2	10
1 deputy do.	0	1	8
1 gunsmith	0	2	2

Fort William.

	£	s.	d.
1 storekeeper	0	4	0
1 gunsmith and servant	0	2	0
1 wheelwright and servant	0	2	0

1708. 30 Dec. Ghent and Bruges retaken by the allies; the Duke employed 150 pieces of ordnance in the siege of Ghent.

1709. 16 Jan. Her majesty's warrant directing powder, shot, ammunition, and other stores should be forthwith sent to Port Mahon, with officers, &c., as follows, with their pay per diem :—

	£	s.	d.
2 engineers at 10s. each	1	0	0
3 fireworkers at 4s. each	0	12	0
Firemaster and master gunner	0	10	0
9 bombardiers at 2s. each...	0	18	0
20 gunners at 1s. 6d. each	1	10	0
10 miners at 1s. 6d. each	0	15	0
60 mattrosses at 1s. each	3	0	0
Surgeon	0	5	0
Storekeeper and paymaster	0	8	0
Clerk of Stores	0	4	0
4 conductors at 2s. 6d. each	0	10	0
4 carpenters at 3s. each	0	12	0
4 smiths at 3s. each	0	12	0
2 wheelwrights at 3s. each	0	6	0
3 coopers at 3s. each	0	9	0
4 masons at 3s. each	0	12	0
1 turner	0	3	0
1 plumber	0	3	0

4 Feb. The House voted £180,000 for the charge of the Office of Ordnance.

16 Feb. Her majesty's warrant directing the train of artillery under the Earl of Galloway in Spain should be augmented, and "that you will cause the carriages, powder, shot, ammunition and stores to be sent to Spain accordingly."

Colonel Michael Richards commander in chief of the British artillery in Spain.

Feb. Colonel Borgarde, after regulating the ordnance in Minorca, returned with his detachment to Catalonia, and was at the sieges of Balaguer and Villa Nova as officer under Colonel Borgarde.

Captain ....Richard Silver.
Gentleman of ordnance ....Thomas Pattison.
Fireworkers .... { Abraham Carpenter. John Thornton. Richard Somerfield.
Lieut.-colonel and comptroller ......George Hay.
1 sergeant
2 corporals
7 bombardiers
12 gunners
20 mattrosses
1 drummer

1709. 17 June. The British troops and artillery marched from Ghent and joined the allied army near Menin on the 21st.

27 June. The allies invest the town and citadel of Tournay with 104 cannon, 24 mortars, and 42 pontoons; the town capitulated the 28th July; and the citadel the 31st August.

11 Sept. The battle of Malplacquet gained by the allies.

23 Sept. The siege of Mons commenced; it capitulated on the 23rd of October; Colonel Hara, of the British artillery, was wounded at the siege.

31 Oct. One bombardier, 11 gunners, 15 mattrosses join Colonel Richards at Tarragona from England.

Fireworkers, { P. Donking Thomas Bassett H. Humphries
6 bombardiers
2 carpenters belonging to bombs, paid by Colonel Richards at Barsalonia.

1710. 8 Jan. The House voted £130,000 for the Office of Ordnance, and the sum of £154,324 15s. 8d. for the payment of debts.

25 Feb. Lieutenant-Colonel George Hay appointed comptroller and lieutenant-colonel under Colonel Michael Richards.

15 April. The allied army took the field; their artillery consisted of 262 pieces of cannon, 20 mortars and howitzers, and 40 pontoons.

21 April. They entered the French lines at Pont-à-Vendin.

1710. 23 April. Commenced the siege of Douay with 200 pieces of cannon, eighty of them 24-pounders; it surrendered the 28th June; 40 pieces of brass cannon, 200 iron cannon, 8 mortars, with ammunition and small arms, were found in the place.

6 July. The artillery under Colonel Richards were engaged in the battle of Almenara; they afterwards besieged Castle Moreon.

15 July. The allies under the Duke besiege Bethune, which surrendered the 28th August; also St. Venant and Aire, which ended the campaign.

1 July. Lieutenant-Colonel Alexander Hara appointed chief fireworker.

John Baxter, mate to ditto.

Major Jonas Watson, chief bombardier.

George Musgrave, appointed chief petardier.

The same number of bombardiers, petardiers, and gunners as for several years.

Annapolis Royal (formerly Port Royal, on the Coast of Nova Scotia) taken by Colonel Nicholson with an expedition of 2000 men and artillery.

30 July. Her majesty's warrant directing an establishment of artillery for the security of Annapolis Royal as follows, with their pay per diem :—

	£	s.	d.
1 captain	0	10	0
1 lieutenant	0	4	0
Surgeon	0	5	0
2 sergeants at 1s. 8d. each	0	3	4
2 corporals at 1s. 4d. each	0	2	8
11 gunners at 1s. 4d. each	0	14	8
40 mattrosses at 1s. each	2	0	0
Engineer	0	10	0
Storekeeper and fireworker	0	6	0
3 bombardiers at 2s. each...	0	6	0
2 armourers at 3s. each	0	6	0

3 Aug. Warrant directing that in making up the accounts of William Hubbald, gent., deceased, late paymaster of the train of artillery, A.D. 1689 for the reducing of Ireland, allowance be made of several sums by him paid in Ireland, amounting to £55,418 1s. 1¼d., and also £8,437 0s. 5¾d. paid here. Signed by the Lord Treasurer Godolphin.—*Harl. M.* 2264, 195.

1710. 13 Aug. Colonel M. Richards commanding the artillery in Spain.

Colonel	Borgarde	6 sergeants
Major	Davis	6 corporals
Comptroller	Lieut.-Colonel Hay	1 bombardier
Paymaster	J. Jeffeies	38 gunners
Adjutant	James Richards	87 mattrosses
Quartermaster	Thomas Lightfoot	2 drummers
Waggon master	James Burton	2 surgeon's mates
Surgeon	John Scott	3 engineers
Captains	Silver Bradbury	1 clerk of stores 8 conductors
Lieutenants	Pattison Hull	4 wheelwrights 5 smiths
Fireworker	Burnet Godfrey	1 gunsmith
Gentlemen of Ordnance	J. Lewis William Allen	2 coopers 1 collar-maker
		4 carpenters
		1 provost marshall
		2 assistants

20 Aug. Colonel Borgarde and his detachment were in the battle at Saragossa; they captured the greatest part of the enemy's artillery.

The British artillery lost 80 men killed and wounded, and 300 artillery mules hamstrung; Colonel Borgarde received four wounds.

Dec. They fought a battle at Villa Veciosa, lost their artillery, and retreated into the kingdom of Arragon; Colonel Borgarde wounded by a cannon shot in the leg and taken prisoner.

18 Sept. A commission appointing Sir Thomas Parker, knt., chief justice of the Queen's Bench, Thomas Erle, Esq., lt.-general of the ordnance, and several others, commissioners for putting in execution the Act for the better fortifying and securing the harbours and docks at Portsmouth, Chatham, and Harwich. Signed by all the Lords of the Treasury.—*Harl. M.* 2264, 217.

29 Nov. Grant unto James Pendlebury, esq., of the office of master gunner within Great Britain and elsewhere, with the wages of 2s. by the day. Signed by the four first Lords of the Treasury.
Harl. M. 2264, 245.

Dec. List of officers, ministers, and attendants of her majesty's royal artillery in the Low Countries, with their pay per diem:—

		£	s.	d.
Colonels	J. H. Hopkey	1	5	0
	J. Pendleburgh	1	5	0
Lieut. colonels	A. Hara	1	0	0
	J. Watson	1	0	0

		£	s.	d.
Major	William Bousfield	0	15	0
Paymaster	William Leathes, and Clerk	0	13	6
Adjutant	T. Ardes	0	6	0
Quartermaster	Godfry Franks	0	6	0
Chaplain	Josias Sanby	0	6	8
Commissary of horse	C. Bates	0	6	0
Master surgeon	T. Paulet	0	6	8
Mate	J. Ellis	0	3	0
Kettle drummer	T. Kennedy	0	3	0
Driver to do.	T. Wright	0	1	6
Provost	T. Hoyd	0	2	6

1st Company.

		£	s.	d.
Captain	R. Guyhen	0	10	0
Lieutenant	P. Gilmoyden	0	6	0
Gentlemen	William Allen	0	4	0
	T. Forbes	0	4	0
3 sergeants at 2s. 6d. each		0	7	6
3 corporals at 2s. each		0	6	0
28 gunners at 1s. 6d. each		2	2	0
44 mattrosses at 1s. each		2	4	0
8 pioneers at 1s. each		0	8	0

2nd Company.

		£	s.	d.
Captain	Christopher Briscoe	0	10	0
Lieutenant	T. Ardes	0	6	0
Gentlemen	Thomas Holman	0	4	0
	J. Edgerton	0	4	0
3 sergeants at 2s. 6d. each		0	7	6
3 corporals at 2s. each		0	6	0
28 gunners at 1s. 6d. each		2	2	0
44 mattrosses at 1s. each		2	4	0
8 pioneers at 1s. each		0	8	0

3rd Company.

		£	s.	d.
Captain	A. Bennet	0	10	0
Lieutenant	E. French	0	6	0
Gentlemen	G. Ackenhead	0	4	0
	T. Brooks	0	4	0
3 sergeants at 2s. 6d. each		0	7	6
3 corporals at 2s. each		0	6	0
28 gunners at 1s. 6d. each		2	2	0
44 mattrosses at 1s. each		2	4	0
8 pioneers at 1s. each		0	8	0
4 bombardiers at 2s. each		0	8	0
Commissary of stores	Thos. Marwood	0	8	0
Assistant do.	T. Casgaoine	0	4	0
Clerk of stores	Thos. Jones	0	4	0
	T. Washington	0	4	0

	£	s.	d.
11 conductors at 2s. each	1	2	0
Conductor and cooper, T. Parker	0	3	0
Master carpenter	0	4	0
His mate	0	2	6
5 carpenters at 2s. 6d. each	0	12	6
Master wheelwright	0	4	0
1 assistant	0	2	6
Master smith	0	4	0
3 smiths at 2s. 6d. each	0	7	6
Master tinman	0	4	0
1 tinman	0	2	6
Bridge-master, James Deal	0	5	0
2 carpenters at 2s. each	0	4	0
20 pontoon men at 1s. 6d. each	1	10	0

1711. 1 Feb. Richard King appointed colonel to the train of artillery ordered on an expedition.

George Spencer appointed major to the train of artillery ordered on an expedition.

10 Feb. John Hopkey appointed first colonel to the train in Flanders.

James Pendlebury appointed second colonel and comptroller to the train in Flanders.

Jonas Watson appointed first lieutenant-colonel to the train in Flanders.

Alexander Hara, lieutenant-colonel ditto.

James Petite, major ditto.

12 Feb. Colonel Richard King ordered on an expedition with the following ordnance, arms, officers, non-commissioned officers, gunners, mattrosses, &c. :—

Iron ordnance on travelling carriages with limber complete	24-pounders	12
	12 do.	8
Mortar pieces, $7\frac{3}{4}$ inch		4
Petards		4

List of officers and ministers for this train :—

		Per diem. £	s.	d.
Colonel	Richard King	1	5	0
Major	George Spencer	0	15	0
Adjutant		0	6	0
2 engineers, each 10s.		1	0	0
Commissary and paymaster		0	8	0
His clerk		0	4	0
2 conductors, each 2s. 6d.		0	5	0
3 labourers, each 2s.		0	6	0
Captain		0	10	0
Lieutenant		0	6	0

	£	s.	d.
2 gentlemen of the ordnance, each 4s.	0	8	0
Surgeon	0	5	0
3 sergeants, each 2s. 6d.	0	7	6
3 corporals, each 2s.	0	6	0
20 gunners, each 1s. 6d.	1	10	0
40 mattrosses, each 1s.	2	0	0
2 carpenters, each 3s.	0	6	0
2 smiths, each 3s.	0	6	0
2 wheelwrights, each 3s.	0	6	0
Collar-maker	0	3	0
Cooper	0	3	0
2 fireworkers, each 4s.	0	8	0
8 bombardiers, each 2s.	0	16	0

Additional MS. 5795.

1711. 30 April. The allied army assembled under the Duke at Orchies, their artillery consisting of 111 pieces of cannon, 8 howitzers, and 40 pontoons.

May. An expedition under Brigadier Hill sailed from England for the attack of Quebec, which failed.

List of officers to attend the train on this expedition, with their pay per diem :—

	£	s.	d.
Colonel	1	5	0
Major	0	15	0
Adjutant	0	6	0
2 engineers at 10s. each	1	0	0
Paymaster	0	10	0
Clerk of stores	0	4	0
2 conductors at 2s. 6d. each	0	5	0
2 labourers at 2s. each	0	4	0
Captain	0	10	0
Lieutenant...	0	6	0
2 gentlemen of the ordnance at 4s. each	0	8	0
Surgeon	0	5	0
His mate	0	3	0
3 sergeants at 2s. 6d. each	0	7	6
3 corporals at 2s. each	0	6	0
20 gunners at 1s. 6d. each	1	10	0
40 mattrosses at 1s. each	2	0	0
8 bombardiers at 2s. each	0	16	0
2 carpenters at 3s. each	0	6	0
2 smiths at 3s. each	0	6	0
2 wheelwrights at 3s. each	0	6	0
1 collar-maker	0	3	0
1 cooper	0	3	0
2 fireworkers at 4s. each	0	8	0

1711. 3 Aug. The Duke determined to attempt getting within the enemy's lines, which began at Bouchain on the Scheld and were continued along the Senset and the Scarpe to Arras, &c., entrenched and fortified by a large ditch, defended with redoubts and other works.

4 Aug. Brigadier Sutton was sent with the artillery and pontoons to make bridges over the Scarpe.

5 Aug. The whole army, after ten hours' march, completely succeeded in their attempt. Brigadier Sutton sent to England with the news.

10 Aug. Bouchain and Wavrechain invested; Colonel Armstrong, who was acting as quartermaster-general, showed great abilities by completing in one night a surprising redoubt with double fausse, and mounting 24 large pieces of cannon with the British standard of our train flying, and saluting the enemy in their works in Wavrechain with several rounds of his cannon at daylight.[1]

Colonel Borgarde returned on parole to England for the cure of his wound, and was exchanged for a colonel of the enemy.

Estimate of Ordnance land service, £130,000.

1712. Richard, Earl of Rivers, appointed master-general; he died in August.

1 Oct. James, Duke of Hamilton and Brandon, succeeded to the master-generalship.

7 June. The allied army besiege Quesney, which surrendered the 4th July. The English army separate from the allies, the Duke of Ormond sending a train of artillery and ammunition to Dunkirk.

Oct. Colonel Borgarde appointed firemaster of England.
Colonel M. Richards appointed chief engineer.
Talbot Edwards, second engineer.
Colonel Christopher Lilly, third engineer.
Thomas Phillips, fourth engineer.

1 July. Brigadier John Hill appointed lieutenant-general of the ordnance.

---

1 "The standard of our train *flying and saluting*."—In the campaigns in Flanders the head quarters of the army was always in the artillery camp. On the line of march the British standard was carried flying on one of the guns, a socket for the purpose being fitted on the trail. While in camp the standard was always hoisted in the artillery camp. Salutes should always be fired from the batteries which fly the national flag, and the flags of palaces and fortifications have, from time immemorial, always been in the custody of the artillery as represented by their master gunners, and the artillery alone should hoist and haul down the colours. See also the detail of the order of march at Marlborough's funeral (pp. 201-2).

1712. 7 July. Brigadier Hill with English forces take possession of Dunkirk.

1712. Ordnance estimate for 1712 :—

	£	s.	d.
For the charge of the Office of Ordnance for land service	111,983	10	4
Carrying on and finishing the fortification of Edinburgh Castle	2,500	0	0
For the fortification of Dunbarton Castle	1,620	0	0
do. do. Fort William	308	6	9
	£116,411	17	1

1713. Ordnance estimate for 1713 :—

Charge of the Ordinary of the Office of Ordnance	28,273	13	9
Pay of the Officers of the Train in Flanders from the 16 April to the 23rd June, 1713, and for the charge of bringing home the stores	3,428	6	0
Purchasing two hundred tons of Saltpetre for supply of the stores	9,000	0	0
Charge of an Engineer and Storekeeper at Jamaica from 1st April to 30th September, 1713	228	5	0
Charge of an Engineer at New York for the year 1713	182	10	0
Officers of the Train in Spain from 1st April to 30th September, 1713	5,220	1	6
Charge of the Officer of Ordnance at Port Mahon for one year	4,544	5	0
Do. do. at Gibraltar	3,631	15	0
Do. do. at Annapolis Royal	2,162	12	6
Stores sent to Placentia	5,473	10	11
Charge of an Engineer, Storekeeper and Gunners for Placentia for one year	1,076	15	0
Charge of the Officers belonging to the Artillery in North Britain for one year	1,475	18	9
	£63,697	13	5

1714. Ordnance estimate for 1714 :—

Charge of the Office of Ordnance for the land service	55,281	16	0
For the Military Officers and Chaplains that served in the Train of Artillery in Flanders and Spain, and on several expeditions, which, with the allowance they have on the Establishment in the Office of Ordnance, is to complete their half-pay for the year 1714	2,188	9	2
	£57,470	5	2

1712. 15 Aug. John Jeffreys appointed lieutenant-colonel and comptroller to the artillery under Colonel Michael Richards.

1713. 6 May. Her majesty's warrant directing a proportion of ordnance, carriages, ammunition, and stores, with an establishment of officers, &c., for the Garrison of Placentia, with their pay per diem.

Iron ordnance	32-pounders	10
	18 do.	8
	9 do.	8
	3 do.	4
Brass	7-inch mortars	2
	Cohorn do.	28
Iron	do. do.	12

	£	s.	d.
Engineer	1	0	0
Master gunner	0	5	0
20 gunners at 1s. 6d. each	1	10	0
Carpenter	0	4	0
Mason	0	4	0
Smith	0	3	0
Armourer	0	3	0

23 June. Her majesty's warrant directing the officers, ministers, gunners, &c., belonging to the ordnance mentioned in the annexed list, should be continued at Dunkirque for our service:—

	Per diem. £	s.	d.
Lieutenant-colonel	1	0	0
Surgeon	0	3	0
Captain	0	10	0
Lieutenant	0	6	0
3 sergeants at 2s. 6d. each	0	7	6
4 corporals at 2s. each	0	8	0
1 bombardier at 2s.	0	2	0
30 gunners at 1s. 6d. each	2	5	0
60 mattrosses at 1s. each	3	0	0
Commissary of stores	0	8	0
Conductor	0	2	6
Conductor and cooper	0	3	0
2 wheelers at 3s. each	0	6	0
1 smith	0	3	0

24 June. General Hill presented to the House of Commons an account of half-pay for military officers and chaplains that had served in the train of artillery of Flanders, Spain, and on several expeditions, &c., which was referred to the Grand Committee of the Supply.

1714. 1 June. The House of Commons grant £2,188 9s. 2d. for the military officers and chaplains that served in the train of artillery in Flanders and in Spain, and on several expeditions, which, with the allowance they have on the establishment in the Office of Ordnance, is to complete their half-pay for the year 1714.

1714. Oct. John, Duke of Marlborough, appointed master-general.

Thomas Earle appointed lieutenant-general.

1715. Colonel Michael Richards appointed surveyor-general.

Jan. Colonel John Armstrong appointed chief engineer.

Edward Ashe, clerk of the ordnance.

James Craggs, clerk of the deliveries.

Harry Mordant, treasurer.

Thomas Frankland, clerk to the master-general.

4 Aug. His majesty's warrant to John, Duke of Marlborough, requiring him to have a train of brass ordnance, mortars, ammunition, &c., with officers, gunners, bombardiers and others, in readiness to attend our land forces as we shall order and appoint in consequence of the rebellion.

Kettle drum, complete		1
Brass	12-pounders	2
	6 do.	8
	Howitzers	2
	3-pounders	18

	Per diem. £	s.	d.
Colonel	1	5	0
Lieutenant-colonel	1	0	0
Major	0	15	0
Comptroller	1	0	0
Paymaster	0	10	0
Clerk to do.	0	3	6
Adjutant	0	6	0
Quartermaster	0	6	0
Commissary of horse	0	6	0
Waggon-master	0	6	0
4 conductors at 2s. 6d. each	0	10	0
Surgeon	0	6	0
do. mate	0	3	0
Kettle drummer	0	3	0
do. driver	0	1	6
Provost marshall	0	2	6
2 provosts at 1s. 6d. each	0	3	0

	Per diem. £	s.	d.
2 captains at 10s. each	1	0	0
2 lieutenants at 6s. each	0	12	0
4 gentlemen of ordnance at 4s. each	0	16	0
6 sergeants at 2s. 6d. each	0	15	0
6 corporals at 2s. each	0	12	0
4 drummers at 1s. 6d. each	0	6	0
2 fireworkers at 4s. each	0	8	0
4 bombardiers at 2s. each...	0	8	0
60 gunners at 1s. 6d. each	4	10	0
120 mattrosses at 1s. each	6	0	0
Commissary of stores	0	8	0
Clerk to do.	0	4	0
6 conductors at 2s. 6d. each	0	15	0
Master carpenter	0	4	0
2 mates at 3s. each	0	6	0
Master wheelwright	0	4	0
2 mates at 3s. each	0	6	0
Master smith	0	4	0
2 mates at 3s. each	0	6	0
2 coopers at 3s. each	0	6	0
Master collar-maker	0	4	0
Mate to do.	0	3	0
Bridge-master	0	5	0
2 corporals at 2s. each	0	4	0
Tinman	0	2	6
20 pontoon men at 1s. 6d. each	1	10	0

1715. 27 Nov. His majesty's warrant requiring the master-general to have in readiness and embarked in proper store ships, for an expedition, several pieces of ordnance with ammunition and stores, officers, gunners, bombardiers, and other attendants, with their pay per diem:—

Brass ordnance	18-pounders	6
	12 do.	8
	9 do.	2
	Howitzers	4
	7-inch mortars	4
	Cohorn	40
	Petards	4

		£	s.	d.
Colonel	Albert Borgard	1	5	0
Surgeon	John Pawlett ...	0	6	8
Assistant	Samuel Marshall	0	3	0
Commissary of stores and paymaster	H. Hutcheson ...	0	8	0
His clerk	Joseph Burton ...	0	4	0
6 conductors at 2s. 6d. each ...	... ...	0	15	0
Chief engineer	... ...	1	0	0
6 engineers at 10s. each	... ...	3	0	0

	£	s.	d.
2 captains at 10s. each { Zach. Smith, James Richards }	1	0	0
2 lieutenants at 6s. each { James Deal, Gadby Franks }	0	12	0
2 gentlemen of ordnance at 4s. each { Jonathan Lewis, Anthony Brown }	0	8	0
4 fireworkers at 4s. each......... { Thomas Bassett, John Melledge, Abm. Carpenter, Geo. Michaelson }	0	16	0
8 bombardiers at 2s. each	0	16	0
5 sergeants at 2s. 6d. each	0	12	6
5 corporals at 2s. each	0	10	0
30 gunners at 1s. 6d. each	2	5	0
50 mattrosses at 1s. each	2	10	0
2 drummers at 1s. 6d. each	0	3	0
Master carpenter	0	4	0
4 carpenters at 3s. each	0	12	0
Master wheeler	0	4	0
3 wheelers at 3s. each	0	9	0
Master smith	0	4	0
Master collar-maker	0	4	0
Collar-maker	0	3	0
2 coopers at 3s. each	0	6	0
*Captain of woolpacks and fascines, McNeal	0	6	0
4 smiths at 3s. each	0	12	0
Sergeant of pontoons	0	2	6
6 pontooners at 1s. 6d. each	0	9	0

1715. Colonel Borgard, with his train, arrived at Leith the 18th January, and joined the Duke of Argyle at Stirling the following day. The army marched towards Perth.

6 Dec. Captain Zach. Smith received clothing at the Tower for his company, consisting of

3 sergeants
2 corporals
4 bombardiers
15 gunners
2 pontoon men
25 mattrosses
1 drummer

1716. The Board of Ordnance ordered Colonel Borgarde to lay before them tables and draughts of all natures of brass and iron cannon, mortars, &c., which was done accordingly and approved of. After the said draughts had been presented, two 24-pounder brass cannon were ordered to be cast by Mr. Bayley in his foundry at Windmill Hill, at the casting of which Colonel Borgarde was ordered to be present.

* This officer—the captain of the woolpacks and fascines—appears for the first time.

In the founding the metal of one of the guns blew into the air, burnt many of the spectators, of which seventeen died out of twenty-five persons; Colonel Borgarde received four wounds.

1716. 30 April. In the muster taken at Blackheath of the two companies of artillery which had served on the expedition to Scotland, the senior sergeant had served in Flanders 25 years, senior corporal 22 years ditto, senior bombardier ditto, most of the gunners and mattrosses 10 and 12 years ditto, also in Spain and bomb vessels.

GEORGE R.

1715. 30 July. Right trusty and right intirely beloved cousin and councellor, we greet you well. Whereas King Charles the Second, and King James the Second, did make an establishment of the Office of Ordnance, in a book intituled Instructions for the Government of the Office of our Ordnance under our Master-General thereof, committed to five principal officers, and signed by the said King Charles and King James, and the principal secretarys of state, which were confirmed by His Majesty King William, and Her late Majesty Queen Anne: our will and pleasure, and express command is, That all orders, warrants, and instructions or directions therein contained and given by the said King Charles and King James as aforesaid, shall continue in force to be obeyed and observed by all and every person or persons who in our said office are or shall be concerned, as fully and amply as the same was, or ought to be observed, in the reigns of the said King Charles, King James, King William, and Queen Anne, and for your so doing this shall be your sufficient warrant untill our further pleasure be known. Given at our Court at St. James's this thirtieth day of July, 1715, in the first year of our reign.

By His Majesty's command,

JAMES STANHOPE.

To John, Duke of Marlborough, our Master-General of our Ordnance.

13 Sept. The lieutenant-general and surveyor-general direct that Major Bousfield, as mate to the master-gunner, have the care and command of the two companies of gunners and mattrosses quartered at Deptford and Woolwich.

That every Saturday at 3 o'clock in the afternoon they be mustered and paid at the gun shedd on Blackheath, and that the messenger do not fail of being precisely there at the time.

That such of the said companies as shall from time to time be employed at the Tower be mustered and paid there.

That Colonel Borgarde do order them to be relieved from time to time as the service shall require.

1715. 4 Oct. That Mr. Thomas White agree for rebuiling the butt at Blackheath, not exceeding £15, and that the messenger pay 1d. a shot for what have been or shall be taken out of it, upon Mr. White certifying the same.

1715. Half-pay of the officers and chaplain who have served well in the train of artillery in Flanders, Spain, and on several expeditions:—

		What will make up their salary half-pay. £	s.	d.
Colonels	John Henry Hopkey	28	2	6
	James Pendlebury	1	12	6
	Richard King	128	2	6
Lieut.-colonels	Jonas Watson	127	15	0
	Roger Davis	182	10	0
	John Jeffreys	182	10	0
Majors	Lieut.-Col. Duterne	136	17	6
	James Petite	136	17	6
	George Spencer	56	17	6
	Francis Hawkins	41	5	0
Engineers	John Redknapp	91	5	0
	Maximilian Faviere	91	5	0
	Jean Chardellon	91	5	0
	John Greul	91	5	0
	John Romer	73	0	0
Sub-engineers	Bloom Williams	50	0	0
	Francis Seys	50	0	0
	Nicholas Dubois	50	0	0
	Edward Ridley	50	0	0
	Noel Marchand	50	0	0
Captains	Peter Gylmoiden	54	15	0
	Thomas Marwood	54	15	0
	James Richards	73	0	0
	Thomas Pattison	73	0	0
	Zachariah Smith	51	5	0
Firemaster	Thomas Heydon	33	0	0
Lieutenants	Godfrey Franks	18	5	0
	Thomas Holman	18	5	0
	Francis Hull	36	10	0
	Henry Piller	36	10	0
	Thomas Hughes	18	5	0
	James Deale	18	5	0
	Peter Stepkins	36	10	0
	John Roope	36	10	0
	Joseph Burton	36	10	0
Adjutants	George Minnens	36	10	0
	James Gernon	54	15	0
Quartermasters	George Aikenhead	36	10	0
	Jonathan Lewis	36	10	0
	Thomas Salisbury	18	5	0

		What will make up their salary half-pay. £	s.	d.
Gentlemen of ordnance	John Brooks	18	5	0
	Henry Garnett	18	5	0
	Richard Sumerfield	36	10	0
Master-gunner from Barbadoes	James Fawcett	45	12	6
Fireworkers	Abraham Carpenter	18	5	0
	Burnet Godfrey	18	5	0
	Humphry Hutcheson ...	18	5	0
	Edward Backhouse	18	5	0
	Michael Crofts...	18	5	0
	Thomas Basset	18	5	0
	John Millidge...	18	5	0
	John Humphreys	18	5	0
	John Cartwright	18	5	0
Chaplain	Josiah Sandby	60	16	8
		£2,832	8	6

1715. Ordnance estimate for 1715 :—

	£	s.	d.
Allowance and wages to the master-general, principal officers, storekeeper, engineers, clerks, labourers, gunners, firemasters, fireworkers, petardiers, bombardiers, &c., daily attendant and employed in the office at the Tower, Berwick, Carlisle, Chatham, Chester, Gravesend, Guernsey, Hull, Jersey, Plymouth, Portsmouth, Sheerness, Tilbury, Tinmouth, Upnor Castle, Windsor, &c., with the rent of houses for officers, governors, storekeepers, clerks, labourers, &c., and of storehouses in Goodman's Fields, Plymouth, Portsmouth, &c.	13,235	7	6
The incident charges of the offices in taking remains of stores, paying extra labourers, books, paper, coals, candles, carriages, messages, postage of letters at the Tower and other garrisons, and abroad..	6,600	0	0
Repairing of platforms, making of new and repairing of old carriages at the several forts, castles, and garrisons, in Great Britain, Guernsey and Jersey	4,000	0	0
Rebuilding and repairing of storehouses, barracks, bridges, gates, pallisades, &c., at the said forts, castles, and garrisons..	10,000	0	0
For furnishing beds, bedsteads, sheets, &c., and repairing the same, at the said forts, castles, and garrisons..	1,500	0	0
The expense of shot, powder, ladles, spunges, flaggs, and other small stores for the guards and garrisons in Great Britain..	3,500	0	0

	£	s.	d.
The yearly charge of the offices, &c., at Port Mahon	3,504	0	0
do. do. at Gibraltar	3,175	10	0
do. do. at Annapolis Royal	730	0	0
do. do. at Placentia	1,076	15	0
do. do. in North Britain	1,475	18	9
For stocking and setting up 8,000 new arms, and for stocking, cleaning, and repairing old arms returned since the peace	10,000	0	0
For 10,000 new arms, contracted for and ordered to be brought into the stores	11,000	0	0
For 300 tuns of salt petre, having but 200 tuns in store, whereas there was kept in store during the last peace near 1,000 tuns............................	13,500	0	0
For putting the battering and travelling trains of artillery in a proper condition, of which 110 guns have always been maintained in the out garrisons, viz., Berwick, Chester, Hull, St. James' and Whitehall, Plymouth, Portsmouth, Isle of Wight, and Edinburgh, besides the Royal Artillery in the Tower	5,000	0	
The extra expences incurred upon the demise of her late Majesty, by order of the Regency, at the several forts, castles, in south and north Britain	2,500	0	0
May 31. For half pay (as per list)	2,832	8	6
	£93,629	19	9

MEMORANDUM.

That the fortifications for Port Mahon, Gibraltar, Annapolis, and Placentia are not included in the foregoing estimate.

That the repairing of the fortifications of North Britain, as estimated by the engineer sent thither since the demise of her late majesty, for securing the same from an insult and for lodging the soldiers' amounts to £15,423 10s. 3½d.

1716. Estimate for 1716 :—

	£	s.	d.
The ordinary of the office, comprising salaries, rents, repairs of storehouses, barracks, platforms, carriages, stores for garrisons, and other incident charges	38,835	7	6
The establishments of Mahon, Gibraltar, Annapolis, Placentia, and North Britain	9,962	3	9
The charges of materials for the fortifications, sent to Mahon	2,965	10	0
The charge of £1,000 per month for the continuance of the said fortifications, in pursuance of his majesty's orders, which reckoned for six months in the year 1716, is	6,000	0	
For supplying the stores with arms in the room of those delivered on the occasion of the rebellion...	22,467	15	

	£	s.	d.
The extra charge of repairing fortifications, fitting out a train of artillery, as also of ammunition and other stores of war provided on account of the rebellion	39,660	10	3½
For half pay of the officers and chaplain who have served well in the trains of artillery in Flanders, Spain, and on several expeditions, as per list	2,605	11	8
	£122,496	18	2½

N.B. The explanation of the first article of £38,835 7s. 6d. in this estimate, is in the six first articles of the preceding estimate, and the second article is explained by the five next articles in the same estimate.

The half-pay list the same as last estimate, with the following exceptions of salary altered to:—

		£	s.	d.
Engineer	John Greul	45	12	6
Captains	James Richards	36	10	0
	Zach. Smith	25	12	6
Lieutenants	Godfry Franks	9	2	6
	Francis Hull	left out of the list.		
	James Deal	9	2	6
	Joseph Burton	18	5	0
Quartermaster	Jonathan Lewis	18	5	0
Fireworkers	Abraham Carpenter	9	2	6
	Humphry Hutchenson	9	2	6
	Thomas Bassett	9	2	6
	John Melledge	9	2	6

Proposal for a Regimental Establishment.

1716. 10 Jan. The Board of Ordnance represent to the Duke of Marlborough that, upon examination of the several establishments of the military branch of this office, it appears that above the sum of £11,000 is yearly paid in salaries to persons that are not appointed to any particular service, and are never employed without more than double the said expense, under the titles either of travelling charges or pay.

It is likewise evident, from many years' experience, that the aforesaid establishments are so defective that when there hath been any extraordinary service required this office hath been obliged to inlist other persons, and to make demand in Parliament for payment of the same.

It is therefore humbly proposed, that as vacancyes shall happen in the said establishments that the salarys of them be applied to the forming of one or more companies of gunners, as shall be thought necessary for his majesty's service.

For the better explanation hereof, we humbly lay before your Grace the annexed proposal for a regimental establishment of 4 companies of gunners with proper officers, consisting of 379 persons and 28 engineers, amounting to £15,539 17s. 6d. per annum.

On the other side is an abstract of the several establishments, consisting of 369 persons belonging to the artillery and for 19 engineers, the whole amounting to £16,829 11s. 3d. per annum, by which it appears that what is now humbly offered is not only a less charge, but does consist of more in number.

Out of the number of the 4 companies thus proposed the foreign establishment will not only be supplied, but there will also remain in England near 200 persons at all times ready to march upon any extraordinary occasion, or otherwise to be usefully employed in those garrisons or places where his majesty's chief magazines are kept, of which persons there is now a great want.

Signed Thomas Earle, Edward Ash,
M. Richards, J. Armstrong.

Employments on the North Britain establishment proposed to sink as they become vacant:—

	£	s.	d.
Captain	146	0	0
Engineer	127	15	0
Lieutenant	91	5	0
Commissary	91	5	0
Corporal	22	16	3
10 gunners	182	10	0
6 practitioner gunners	54	15	0
1 petardier	36	10	0
2 miners	54	15	0
Deputy storekeeper at Stirling	30	0	0
Gunsmith	40	0	0
Master smith at Fort William	36	10	0
Wheelwright and servant	36	10	0
6 bombardiers	219	0	0
	1,169	11	3
Difference after the following reduction	32	12	6
	£1,202	3	9

Salarys to be reduced from:—

	£	s.	d.	£	s.	d.
Storekeeper at Edinburgh	111	5	0	91	5	0
Clerk to do.	30	0	0	36	10	0
Armourer	40	0	0	36	10	0
Storekeeper at Fort William	73	0	0	54	15	0
do. at Stirling	52	2	6	54	15	0
	£306	7	6	£273	15	0

An abstract of part of the yearly charge of the military branch of the Office of Ordnance, proposed as they become vacant not to be filled up again :—

		£	s.	d.
A chief engineer	... ...	300	0	0
Second engineer	... ...	250	0	0
Third engineer...	... ...	150	0	0
Extra engineer	... ...	365	0	0
11 engineers ...	... ...	1,100	0	0
4 sub-engineers	... ...	200	0	0
Comptroller of fireworks	... ...	200	0	0
Chief firemaster	... ...	150	0	0
His mate	... ...	80	0	0
8 fireworkers ...	... ...	320	0	0
1 firemaster of ships ...	... ...	60	0	0
Chief petardier	... ...	54	15	0
Chief bombardier	... ...	54	15	0
4 petardiers ...	... ...	146	0	0
24 bombardiers	... ...	876	0	0
Master gunner of Great Britain	... ...	190	0	0
8 mates...	... ...	136	10	0
120 gunners ...	... ...	2,190	0	0
2 captains	... ...	120	0	0
2 first lieutenants	... ...	100	0	0
2 second lieutenants ...	... ...	80	0	0
9 gentlemen of the ordnance...	... ...	360	0	0
30 labourers ...	... ...	780	0	0
North Britain on the other side	... ...	1,202	3	9
The pay of the trains	Gibraltar	2,491	1	6
	Port Mahon	3,248	10	0
	Annapolis...	547	10	0
	Placentia	1,076	15	0
		£16,829	0	3

		£	s.	d.
The following are proposed to continue till further orders:—				
217 master gunners and gunners of garrisons	...	4,605	1	8
Half pay of the officers that served abroad...	...	2,710	15	2
		£24,144	17	1

A regimental establishment humbly offered as more advantageous to the service; per annum :—

	Per diem.				Per annum.		
	£	s.	d.		£	s.	d.
Colonel	1	5	0	...	456	5	0
Lieut.-colonel	1	0	0	...	365	0	0
Major	0	15	0	...	273	15	0
Captain...	0	10	0	...	182	10	0

	Per diem. £ s. d.			Per annum. £ s. d.		
Lieutenants 1st	0	6	0	109	10	0
Lieutenants 2nd	0	5	0	91	5	0
Lieutenant fireworkers 3rd... ...	0	4	0	73	0	0
Lieutenant fireworkers 4th... ...	0	3	0	54	15	0
3 sergeants at 2s. each	0	6	0	109	10	0
3 corporals at 1s. 8d. each	0	5	0	91	5	0
3 bombardiers at 1s. 8d. each... ...	0	5	0	91	5	0
30 gunners at 1s. 4d. each	2	0	0	730	0	0
50 mattrosses at 1s. each	2	10	0	912	10	0
				3,540	10	0
Three companies more				7,336	10	0
				10,877	0	0
1 chief engineer	1	7	6	501	17	0
3 directors at £1 each...	3	0	0	1,095	0	0
6 engineers in ordinary at 12s. each...	3	12	0	1,314	0	0
6 engineers extraordinary at 8s. each	2	8	0	876	0	0
6 sub-engineers at 5s. each	1	10	0	547	10	0
6 practitioner engineers at 3s. each ...	0	18	0	328	10	0
				£15,539	17	0
The remaining articles amounting to ... ... are not included for the following reasons:—				7,315	16	10
The garrison gunners, although appointed and warranted by the master of the ordnance, are as yet paid by the paymaster of the army				4,605	1	8
The half pay of the officers will sink as vacancies happen				2,710	15	2

Granted annually by Parliament.

Old establishment for Mahon.

	Per diem. s.	d.	Per annum. £	s.	d.
1 engineer	20	0	365	0	0
1 sub-engineer ...	10	0	182	10	0
1 storekeeper... ...	8	0	146	0	0
1 clerk of works ...	10	0	182	10	0
1 clerk of stores ...	4	0	73	0	0
1 barrack master ...	5	0	91	5	0
1 firemaster and master gunner	10	0	182	10	0
8 bombardiers ...	2	0	292	0	0
20 gunners	1	6	547	10	0
40 mattrosses	1	0	730	0	0
3 carpenters	3	0	164	5	0
2 smiths	3	0	109	10	0
2 wheelers	3	0	109	10	0
2 coopers	3	0	109	10	0
1 turner	3	0	54	15	0
1 plumber	3	0	54	15	0
2 armourers	3	0	109	10	0
			3,504	0	0

Old establishment for Gibraltar.

	Per diem. s.	d.	Per annum. £	s.	d.
1 engineer	20	0	365	0	0
1 sub-engineer ...	10	0	182	10	0
1 storekeeper & paymaster	—		219	0	0
1 clerk of works ...	10	0	182	10	0
1 barrack master ...	5	0	91	5	0
1 captain	10	0	182	10	0
1 lieutenant ...	6	0	109	10	0
1 fireworker ...	4	0	73	0	0
6 bombardiers ...	2	0	219	0	0
2 sergeants, each ...	2	6	91	5	0
2 corporals, " ...	2	0	73	0	0
20 gunners " ...	1	6	547	10	0
25 mattrosses " ...	1	0	456	5	0
1 smith	4	0	73	0	0
1 cooper	4	0	73	0	0
1 turner	4	0	73	0	0
1 wheeler	3	0	54	15	0
2 armourers ...	3	0	109	10	0
			3,175	10	0

1716. 26 May. His majesty's warrant. The Duke of Marlborough having represented the inconvenience and defects of the present establishment of the military branch of our said Ordnance, amounting to £16,829 11s. 3d., and therewith a scheme shewing that a greater number of gunners, engineers, and other proper officers may be maintained for less than the present expense; and whereas by our warrant of the 27th November, 1715, two companies of gunners and mattrosses were raised for the service of our artillery sent upon the late expedition to North Britain, and having been found always necessary that a sufficient number of gunners with proper officers should be maintained and kept ready for our service; and whereas there are several salaries now vacant of the present old establishment which are not useful, and that other saving may be made by which part of the two companies may be maintained, viz., 1 sergeant, 3 corporals, 3 bombardiers, 30 gunners, and 32 mattrosses, such as have served well abroad during the late wars; and as other salaries shall become vacant in the military branch, you will complete the pay of the rest of the officers, &c., according to the annexed list.

Establishment of two companies:—

First Company.

	Per diem. £	s.	d.
Captain	0	10	0
Lieutenants 1	0	6	0
Lieutenants 2	0	5	0
Lieutenant fireworkers 3	0	4	0
Lieutenant fireworkers 4	0	3	0
3 sergeants, each	0	2	0
3 corporals "	0	1	8
3 bombardiers, each	0	1	8
30 gunners "	0	1	4
50 mattrosses "	0	1	0

Second Company.

The same establishment and pay as the first company.

Explosion at the Gun Foundry at Moorfields.

The original foundery for brass ordnance belonging to Government was in Upper Moorfields in London, near the spot where the chapel erected for the late Rev. J. Wesley now stands, and which, from the circumstance of his having before preached for many years in the foundery itself, is occasionally called by that name. The operation of casting was then, as it still is, an object of some curiosity; and many persons, even of the higher ranks, frequently attended to see the process of running the fluid metal into the moulds. About the year 1716, when Colonel Armstrong was surveyor-general of the ordnance, and George Harrison, Esq., superintendant of the founderies, in which place he had succeeded the former, it was determined to recast the unserviceable cannon which had been taken from the French in the ten successful campaigns of the Duke of Marlborough, and which had hitherto been placed before the foundery and in the adjacent artillery ground. This becoming the more generally known, from the long time the cannon had been publickly exposed, excited a more than common interest; a great number of persons assembled to view the operation, among whom were many of the nobility, general officers, &c., for whoes reception galleries had been prepared near the furnace.

On the same day a native of Schaffhausen, in Switzerland, named Andrew Schalch (who from a common law of his canton, which made it necessary for every person born there to travel for improvement in his profession during three years, had visited different founderies on the Continent, and at length reached England), was attracted to the same place at an early hour, and was suffered minutely to inspect the work then going on. Colonel Armstrong was himself present, when Schalch, being alarmed at some latent dampness which he had observed in the moulds, addressed him in French, and after explaining his reasons for believing that an explosion would accompany the casting of the metal, warned him to retire from the impending danger. The Colonel, who at once comprehended the importance of Schalch's remarks, interrogated him with respect to his knowledge of the art, and found him perfectly conversant with all its principles: he therefore resolved to follow his advice, and quitted the foundery with his own friends and as many of the company as could be prevailed on to believe that danger really existed. Scarcely had they got to a sufficient distance when the furnaces were opened, and the metal rushed into the moulds; the humidity of which, as Schalch had intimated, immediately occasioned a dreadful explosion; the water was converted into steam, and this, by its expansive force, caused the liquid fire to dart out in every direction, so that part of the roof of the building was blown off, and the galleries fell. Most of the workmen were burnt in a dreadful manner, some lives were lost, and many persons had their limbs broken.

A few days afterwards an advertisement appeared in public prints, stating, in substance, that "if the young foreigner who, in a conversation with Colonel Armstrong on the day of the accident at the foundery in Moorfields, had suggested the probability of an explosion from the state of the moulds, would call on the Colonel at the Tower, the interview might conduce to his advantage." Schalch was informed of this intimation by an acquaintance, and he directly waited on Colonel Armstrong, who, after some preliminary discourse, told him that the Board of Ordnance had in contemplation to erect a new foundery at a distance from the metropolis, and that he was authorized, through the representation which he had made of his own conviction of his, Schalch's, ability, to offer him a commission to make choice of any spot within twelve miles of London for the erection of such a building (having proper reference to the extensive nature of the works and carriage of the heavy materials), and also to engage him as superintendent of the whole concern.

This advantageous proposal was readily accepted by Schalch, who immediately began his search for a proper place for the new establishment; and having inspected various spots, he at length fixed on the *Warren* at Woolwich as the most eligible situation. Here the new foundery was erected; and the first specimens of ordnance cast by Schalch were so highly approved, that he was fixed to the situation of master founder, and continued to hold that office for above sixty years, when he retired to Charlton, having been assisted during the latter part of that time by his nephew Lewis Gaschlin. Schalch died in 1776 when about the age of ninety.

Such was the singular train of circumstances that led to the establishment of the Royal Arsenal at Woolwich, and of course to the different institutions that have successively arisen from it.[1]

The Arsenal includes nearly sixty acres, and contains various piles of brick building, among the oldest of which are the foundery and the late Military Academy; these were erected by Sir John Vanbrugh, and have the date 1719 on the upper parts of the leaden pipes that convey the water from the roofs.

The Warren, as this depot was originally called, from its having been previously the site of a rabbit warren, was changed into the Royal Arsenal by his majesty George III.

In Hasted's History of Kent, he mentions, "On the eastern or lower part of the town of Woolwich is the *Gun Yard*, commonly called *the Park* or the *Gun Park*, where there is a great quantity of cannon of all sorts for the ships of war, every ship's guns lying in tiers or rows apart. There is both a civil and military branch of the Office of Ordnance at Woolwich, and under this office, in a place called the *Warren*, artillery of all kinds are cast," &c.

It appears the market-place in Woolwich was changed within the present century. The Gun Wharf formerly occupied the spot where the present market is now held.

1716. 28 March. The expedition which sailed to Scotland in November last returned to England with the exception of:—

1 gentleman of the ordnance......Jonathan Lewis	ordered on the expedition to Inverness.
2 bombardiers	
4 gunners	
4 mattrosses	
2 artificers	

1716. 7 June. His majesty's warrant for a train of artillery "for the campaign in Hyde Park."

Artillery encamped in Hyde Park:—

Brass ordnance on travelling carriages complete	6-pounders	2
	3 do.	8

1 lieutenant................. Godfrey Franks.
2 sergeants.
2 corporals.
20 gunners.

---

1 See an interesting paper on the "Royal Arsenal," by Lieut. Grover, R.E., in the 6th vol. of the Proceedings of the R.A. Institution, in which this version of the origin of the Royal Arsenal is challenged.

1716. 18 Sept. Fees paid to the secretary and clerks of the Ordnance by the members of the newly established train:—

First Company.

		Secretary. £ s. d.		Clerk of the Ordnance. £ s. d.	
Captain	J. Richards ...	2 0 0	...	2 0 0	Total £52 10s.
4 lieutenants ...	Thos. Hughes	2 0 0	...	2 0 0	
	Geo. Aikenhead	2 0 0	...	2 0 0	
	Joseph Egerton	2 0 0	...	2 0 0	
	JonathanLewis	2 0 0	...	2 0 0	
2 corporals, each ...	... ...	1 10 0	...	1 0 0	
1 bombardier	... ...	1 10 0	...	1 0 0	
11 gunners, each ...	... ...	1 0 0	...	0 10 0	
14 mattrosses, each	... ...	0 10 0	...	0 5 0	

Second Company.

Captain	Thomas Pattison	The same proportion as above, which was charged to their accounts.
4 lieutenants ..	Thos. Holman	
	James Deal	
	John Roofe	
	John Brooks	
1 corporal		
2 bombardiers		
11 gunners		
7 mattrosses		

His majesty's warrant, stating that the Duke of Marlborough, master-general of the ordnance, having represented that the number of cannon in our forts, castles, fortifications, &c., are much more than are necessary, or in proportion to what are used on the Continent in like places, was authorized to reduce the number and nature of the said ordnance as proposed in the annexed scheme.

The number of cannon at the several forts, castles, &c., in Great Britain, with a scheme humbly proposed for reducing the same; and also of the flaggs now used, and what they may be reduced to at present.

Names of Places in the several Divisions.	No. of Guns according to the ancient Establishments.	Guns when reduced.	Flaggs at present.	Flaggs when reduced.
Portsmouth Town	401	100	1	1
Blockhouse Fort	32	28	1	...
Southsea Castle	73	43	1	...
Charles's Fort	18	2	1	...
James's "	23	2	1	...
Gosport...	52	...	...	...
The Dock	30	...	...	...
Isle of Wight. Yarmouth	73	13	1	...
Isle of Wight. Sandham	26	12	1	...
Isle of Wight. Carrisbrook	10	7	1	1
Isle of Wight. Cows	20	10	1	...
Hurst Castle	34	18	1	1
Calshott "	25	15	1	...
	817	250	11	3

Names of Places in the several Divisions.	No. of Guns according to the ancient Establishments.	Guns when reduced.	Flaggs at present.	Flaggs when reduced.
Portland Castle...	15	7	1	...
Plymouth & St. Nicholas's Island	311	70	1	1
Pendennis	96	20	1	1
St. Maws	13	10	1	1
Scilly Island	118	30	1	1
	553	137	5	4
Edenburgh	50	40	1	1
Sterling...	32	30	1	1
Fort William	68	30	1	1
Dumbarton	12	12	1	1
Blackness	7	7	1	1
	169	119	5	5
Scarborough Castle	12	8	1	...
Berwick...	76	50	1	1
Holy Island	20	12	1	...
Tinmouth	80	} 20	2	1
Cliffords Fort	30			
Hull	117	50	1	1
Carlisle	31	20	1	1
	366	160	7	4
Sheerness	150	70	1	1
Hawness	27	...	1	...
Gillingham	54	40	1	...
James's Battery	18	...	1	...
Middleton's Battery	10	...	1	...
Upnor	37	21	1	1
Cockham Wood	44	16	1	...
	340	147	7	2
Tilbury	161	60	1	1
Gravesend	17	10	1	...
	178	70	2	1
Dover Castle	45	10	1	1
Moats Bulwark...	13	8	1	...
Arcliff Fort	22	8	1	...
Sandown	18	10	1	...
Deal Castle	29	12	1	1
Walmer	17	10	1	...
Sandgate	16	10	1	...
	160	68	7	2
Guernsey Island	97	80	1	1
Landguard Fort	63	20	1	...
North Yarmouth	17	8	1	...
Chester	96	25	1	1
Jersey Island	149	100	1	1
Windsor	25	25	1	1
Whitehall	6	...	...	...
Tower of London	183	45	1	1
	636	203	7	5
Total	3,219	1,254	51	26

Gibraltar :—

Iron	32-pounders	26	on standing carriages.
	18 "	22	
	12 "	34	
	8 "	1	
	6 "	13	
	5¼ "	2	
	4 "	18	
	3 "	2	
	2 "	2	
Iron	13-inch mortars	5	On their beds.
Brass	10 "	2	
	8 "	4	
	5½ "	12	
	4½ "	14	
Iron	4 " Cohorn	38	

Placentia :—

Iron	18-pounders	8
	9 "	4
Brass Cohorn mortars, small		10

Port Mahon :—

Iron ordnance with their carriages.	42-pounders	8	
	32 "	24	
	18 "	18	
	12 "	12	
	9 "	41	
	8 "	8	
	6 "	26	
	5¼ "		
	4 "	11	
Brass ordnance	18 "	1	
	6 "	1	
	1½ "	6	
Iron mortars	13-inch	4	mounted.
Brass mortars with their beds.	8-inch	5	
	5½ " Cohorns	4	
	4½ " Cohorns	30	

1717. **Estimate for the year 1717 :—**

		£	s.	d.
The ordinary	as in the estimate for 1716.	38,835	7	6
Four foreign garrisons and North Britain		9,962	8	9
For rebuilding of Landguard Fort		2,975	15	9
Building a grand powder magazine at Tilbury		5,800	0	0
Do. do. do. at Sheerness		1,200	0	0
Do. a brass foundry at Woolwich		2,700	0	0
Do. storehouse at Plymouth		3,000	0	0
Do. a barrack for 50 soldiers at Upnor Castle		1,400	0	0
Do. do. for a regiment of foot at Berwick		4,937	10	7
Half pay as per list of 1715, with the exceptions as under		2,266	11	8
		78,077	9	8

In the half-pay list:—

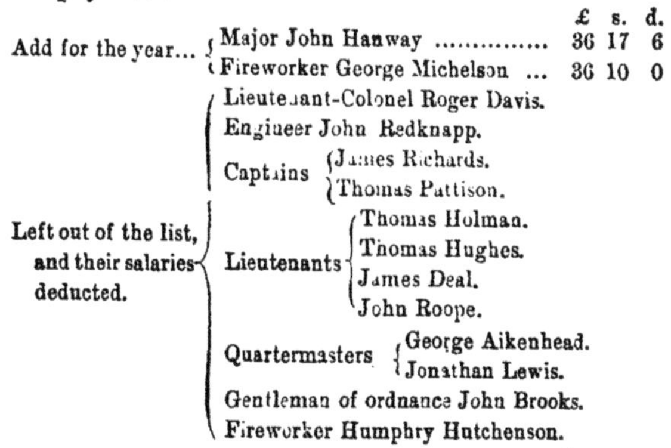

		£	s.	d.
Add for the year...	Major John Hanway	36	17	6
	Fireworker George Michelson ...	36	10	0
Left out of the list, and their salaries deducted.	Lieutenant-Colonel Roger Davis.			
	Engineer John Redknapp.			
	Captains: James Richards. Thomas Pattison.			
	Lieutenants: Thomas Holman. Thomas Hughes. James Deal. John Roope.			
	Quartermasters: George Aikenhead. Jonathan Lewis.			
	Gentleman of ordnance John Brooks.			
	Fireworker Humphry Hutchenson.			

1717. 18 March. Whereas we have thought it necessary for our service to appoint an able and experienced person as superintendant for the inspection and direction of our fortifications and all other military buildings in our kingdom of Great Britain and all other our dominions, and being well satisfied of the long services, great abilities, and experience of our trusty and well-beloved Brigadier-General Michael Richards, master-surveyor of our Ordnance, we do therefore by these presents nominate, constitute, and appoint the said Michael Richards to be superintendant of our fortifications and all other military buildings in our kingdom of Great Britain and all other our dominions; and we do hereby authorise and require you to cause an allowance to be made him, the said Michael Richards, or his assigns, quarterly, by equal portions, by bill and debenture, according to the usage of our said office, the said allowance to commence and to be accounted from the Annunciation of Our Lady next ensuing, and to continue during our pleasure, as also an additional allowance for travelling charges at 20s. per diem, and for so doing this shall be as well to you as to the principal officers of our Ordnance and auditors of our imprests, and all others whom it may concern, a full and sufficient warrant. Given at our Court at St. James's, the 18th day of March, in the fourth year of our reign.

By His Majesty's command,

SUNDERLAND.

To our right trusty, &c., John, Duke of Marlborough, &c.

1717. March. Whereas I have received a warrant from His Majesty dated the 18th instant, signifying His Majesty's pleasure that Michael Richards, Esq., should be superintendant for the inspection and direction of his fortifications and all other military buildings in his kingdom of Great Britain and all other his dominions, as also the additional allowance of 20s. a day for his travelling charges. These are therefore to pray and require you to enter him, the said

Michael Richards, Esq., as His Majesty's superintendant of the fortifications and all other military buildings in Great Britain and other His Majesty's dominions, and to allow him the sum of £500 a year, as also the said additional allowance of 20s. a day for his travelling charges: to commence and to be accounted from the Annunciation of Our Lady next ensuing, and to continue during His Majesty's pleasure, which said allowance of £500 a year is to be paid him, or his assigns, quarterly by equal portions, by bill and debenture, according to the usage of the office, and for so doing this, together with His Majesty's said warrant herewith sent you, shall be your sufficient warrant. Given at the Office of Ordnance, under my hand and seal, the day of March, 1717, in the fourth year of His Majesty's reign.

MARLBOROUGH.

To the Right Honourable Thomas Micklewaite, Esq.,
Lieutenant-General of the Ordnance, &c.

The Board of Ordnance came to a resolution to regulate what was wanting to complete a complete artillery for sea and land service. Colonel Borgard had orders to lay before them draughts of all natures of carriages, wheels, trucks, grape and matted shot, and all sorts of bombs, both great and small, for land and sea service, with a great many other things relating to artillery, too tedious to mention, which the Board approved of.

Colonel Borgard also laid before the Board the ill state of the Laboratory, which they ordered him to put in better order with as little expense as possible, which he accordingly did.

1717. April. Sir Thomas Wheale appointed storekeeper to the Ordnance.

22 Aug. At Hampton Court, present, the King's Most Excellent Majesty in Council.

Upon reading this day at the Board a memorial from the master-general and principal officers of the Ordnance, dated 11th December, 1716, setting forth among other matters that they have prepared an establishment for 29 engineers, at a less expense than the present establishment, and to greater advantage of the service, which very much suffers for want of such a regulation, amounting to £4,297 17s. 6d. per annum. As likewise a new establishment of officers, &c., for Minorca, amounting to £3,756 9s. 2d. per annum; and another new establishment of officers, &c., at Gibraltar, amounting £2,898 14s. 2d. per annum, which they, the said master-general and principal officers, do propose to take place as vacancies shall happen on the three old establishments; and also an account of brass and iron ordnance at Minorca and Gibraltar, with a scheme humbly proposed for reducing the same, great part thereof being unserviceable, and of improper natures.

His Majesty, taking the same into consideration, was pleased, with the advice of his Privy Council, to approve the said three new establishments, and of the scheme for reducing the ordnance, copies whereof are hereunto annexed; and it is hereby ordered the said master-general will cause the necessary directions to be given for putting the same in due and effectual execution accordingly.

Establishment proposed for 29 engineers.

	Per diem.				Per annum.		
	£	s.	d.		£	s.	d.
chief engineer	1	7	6	...	501	17	6
2 directors, each £1	2	0	0	...	730	0	0
2 sub-directors, each 15s.	1	10	0	...	547	10	0
6 engineers ordinary, each 10s. ...	3	0	0	...	1,095	0	0
6 engineers extraordinary, each 6s.	1	16	0	...	657	0	0
6 sub-engineers, each 4s.	1	4	0	...	438	0	0
6 practitioner engineers, each 3s. ...	0	18	0	...	328	10	0
					4,297	17	6

New establishment proposed for Mahon.

	Per diem.		Per annum.		
	s.	d.	£	s.	d.
1 director	20	0	365	0	0
1 engineer extra ...	10	0	182	10	0
1 sub-engineer ...	3	0	54	15	0
1 storekeeper and pay-master	10	0	182	10	0
1 clerk of stores ...	4	0	73	0	0
2 conductors	2	0	73	0	0
1 sworn clerk of the works	3	0	73	0	0
1 do. as barrack master	1	0	(included above)		
1 captain	10	0	182	10	0
1 lieutenant	6	0	109	10	0
1 do.	4	0	73	0	0
2 sergeants, each ...	2	0	73	0	0
2 corporals, „ ...	1	8	60	16	8
5 bombardiers, „ ...	1	8	152	1	8
30 gunners „ ...	1	4	780	0	0
60 mattrosses „ ...	1	0	1,095	0	0
1 master artificer carpenter	4	0	73	0	0
1 smith	3	0	54	15	0
1 wheeler	3	0	54	15	0
1 cooper	2	6	45	12	6
2 other artificers ...	1	4	48	13	4
			3,756	9	2

New establishment proposed for Gibraltar.

	Per diem.		Per annum.		
	s.	d.	£	s.	d.
1 sub-director ...	15	0	273	15	0
1 engineer	6	0	109	10	0
1 storekeeper ...	10	0	182	10	0
1 clerk of stores ...	4	0	73	0	0
2 conductors ...	2	0	73	0	0
1 sworn clerk of works	3	0	73	0	0
1 do. as barrack master	1	0	(included above)		
1 captain	10	0	182	10	0
1 lieutenant ...	6	0	109	10	0
1 do.	4	0	73	0	0
2 sergeants, each ...	2	0	73	0	0
2 corporals „ ...	1	8	60	16	8
4 bombardiers, „ ...	1	8	121	13	4
20 gunners „ ...	1	4	486	13	4
40 mattrosses „ ...	1	0	730	0	0
1 master artificer carpenter	4	0	73	0	0
1 wheeler	3	0	54	15	0
1 smith	3	0	54	15	0
1 cooper	2	6	45	12	6
2 other artificers ...	1	4	48	13	4
			2,898	14	2

NOTE.—See the tables of the old establishments given at page 188.

1718. Establishment of the royal artillery:

1 colonel.
1 lieutenant-colonel.
1 major.
4 companies.
1 adjutant.
1 quartermaster.
1 bridgemaster.

1718. 18 March. Brigadier-General Michael Richards, master-surveyor of our Ordnance, appointed by warrant superintendant-general of his majesty's fortifications, at a salary of £500 per annum.

1 April. Charles Wills, Esq., appointed lieutenant-general of the Ordnance.

25 April. His majesty signified to Colonel Borgarde that he was appointed assistant to the surveyor of the Ordnance at 13s. 4d. per diem.

4 June. Two bomb vessels and two fireships proceeded with the fleet under Sir George Byng to the Mediterranean.

1 July. His majesty's warrant approving of an alteration in the civil branch of the Ordnance.

1719. June. General Wightman, with his forces and a detachment of artillery with four Cohorn mortars, march from Inverness and attack the Spaniards at the pass of Glenshill, who had landed at Kintail the April previous, taking the whole prisoners.

3 July. The lord justice's warrant directing a train of artillery to be got ready to attend his majesty's land forces, ordered to embark upon an expedition to be commanded by the Lord Cobham.

Brass ordnance	24-pounders		4
	9 „		4
	1½ „		6
	10 inch	Mortars	2
	8 „	Mortars	2
	Cohorn	Mortars	30
	Royal	Mortars	12
	Petards		2

		Full pay per diem. £	s.	d.
Colonel	Albert Borgard	1	5	0
Major	William Bousfield, aptd. 1st August	0	15	0
Surgeon	James Barnes	0	6	8
His mate	William Theynne	0	3	4

			Full pay per diem. £	s.	d.
*Captain ...	James Richards				
3 lieutenants	James Deale Jonathan Lewis Peter Stepkins				
4 fireworkers	Burnet Godfry Bennet Smith George Michelson Thomas James	each 4s	0	16	0
3 sergeants					
4 corporals or bombardiers					
20 gunners					
40 mattrosses					
2 drummers					
3 carpenters, each 3s. ...	... ...	... ...	0	9	0
2 wheelers " ...	... ...	... ...	0	6	0
2 smiths " ...	... ...	... ...	0	6	0
1 farrier	... ...	... ...	0	3	0
1 collar-maker	... ...	... ...	0	3	0
1 cooper	... ...	... ...	0	3	0
1 commissary and paymaster	James Burton	... ...	0	10	0
1 clerk of stores	Samuel Gibbs	... ...	0	4	0
5 conductors, each 2s. 6d....	... ...	... ...	0	12	6
4 engineers	John Romer John Selioke Bloom Williams John Hargrave	each 10s.	2	0	0
1 conductor of horses ...	... ...	... ...	0	2	6
12 drivers, each 2s. ...	... ...	... ...	1	4	0
10 watermen, each 2s. 6d.	... ...	... ...	1	5	0
1 cockswain	... ...	... ...	0	2	6

Officers, &c., for two bomb vessels accompanying the expedition:—

Fireworkers	John Winch Edward Backhouse	Speedwell Bomb.
4 bombardiers... ...	... ...	
1 carpenter	... ...	
Fireworkers	Robert Braman Henry Harwood	Furnace Bomb.
4 bombardiers... ...	... ...	
1 carpenter	... ...	

Colonel Borgard was ordered with this expedition to Vigo, which place he "bombarded with 46 great and small mortars of his own projection, which answered their intended end, of which Lord Cobham and

* The pay of the company of gunners was otherwise provided for. This warrant only adds the "staff" necessary on taking the field.

the rest of the general officers can give a better account then himself;" by which bombardment the Castle of Vigo was obliged in the month of October to surrender.

1720. 11 June. His majesty's warrant authorizing an alteration in the establishment of the two companies in England as follows:—

	£	s.	d.
1 captain	0	10	0
Captain lieutenant	0	6	0
1st lieutenant	0	5	0
2nd lieutenant	0	4	0
4 fireworkers, each...	0	3	0
3 sergeants, "	0	2	0
3 corporals, "	0	1	8
12 bombardiers, "	0	1	8
25 gunners, "	0	1	4
5 gunner cadets, "	0	1	4
43 mattrosses, "	0	1	0
5 mattross cadets, "	0	1	0
2 drummers, "	0	1	0

The second company the same:—

	£	s.	d.
1 colonel	1	10	0
1 lieutenant-colonel	1	0	0
1 major	0	15	0

1721. The Honourable Brigadier Michael Richards, surveyor of the Ordnance to his late Majesty King George I., died: his monument may be seen in Charlton Church, Kent.

Quarter book in the Tower:—

Paid their allowance. { Lieutenant-General Charles Wills.
Colonel James Pendlebury, master-general of England.
Colonel H. Hopkey, comptroller of fireworks.

2 Dec. His majesty's warrant for furnishing two bomb vessels with mortars for an expedition by sea:—

13-inch mortars .... { Furnace Bomb 2
Thunder Bomb......... 1

	£	s.	d.
3 fireworkers, each	0	4	0
6 bombardiers, "	0	1	8
2 carpenters "	0	3	0

1721–22. 20 Feb. Colonel John Armstrong appointed master-surveyor of the Ordnance and assistant deputy to the lieutenant-general of the Ordnance.

1721-22. 20 Feb. Whereas we have thought fit by our letters patent under our Great Seal, bearing date the 12th day of this instant February, to constitute and appoint our trusty and well-beloved Colonel John Armstrong, quartermaster-general of our forces, and our chief engineer of Great Britain to be master-surveyor of our ordnance, ammunition, and habiliaments of war within our Tower of London, our Kingdom of Great Britain and Ireland, and all other our dominions, during our pleasure: We do, by these presents, nominate, constitute, and appoint the said John Armstrong to be assistant and deputy to our lieutenant-general of Ordnance for the time being, in such manner as Sir Henry Sheer, Knight, John Charlton, William Bridges, and Michael Richards, Esq., were heretofore nominated, constituted, and appointed, and in the absence of the said lieutenant-general of our Ordnance for the time being to do, execute, and perform all matters to the said office or place of our lieutenant-general of our Ordnance belonging, in as ample manner, to all intents and purposes, as the lieutenant-general of our Ordnance for the time being could or might have done if he were actually present, and as fully and freely as Sir Henry Sheer, Knight, John Charlton, William Bridges, and Michael Richards, Esq., or any former deputys had enjoyed or exercised the said office or deputation. And we have also thought to authorize and require, and we do hereby authorize and require you, to cause an allowance to be made him, the said John Armstrong, out of our treasury of our Ordnance, of the sum of £300 per annum, to be paid him, the said John Armstrong or his assigns, quarterly by equal portions, by bill and debenture, according to the usage of our said office, &c. &c. Given at our Court at St. James's the 20th day of February. 1721-22, in the eighth year of our reign.

By His Majesty's command,

CARTERET.

To the right trusty, &c., John, Duke of Marlborough, &c.

1722. Colonel Borgarde attended the service as formerly at all surveys relating to the artillery, untill the time Colonel Armstrong was made surveyor, after which time, notwithstanding his majesty's signification to him for regulating the artillery for sea and land service, he never was consulted in anything relating to the said service.

1st April. Colonel Borgarde was appointed colonel of the royal regiment of artillery, consisting of four companies.

1st July. William, Earl of Cadogan, appointed master-general.

9th Aug. A train of artillery consisting of 15 pieces of cannon and two mortars, with the great kettle drums belonging to the artillery (on a carriage drawn by two horses), and the two companies of cannoniers and bombardiers, were drawn up in Hyde Park near the gate towards Piccadilly, with other troops to form the procession at the funeral of his grace John, Duke of Marlborough.

The artillery with the two companies of cannoniers and bombardiers, commanded by Colonel Borgarde; and the train consisting of six tunbrils with 12 horses drove by six drivers, three covered waggons by nine horses and three drivers, seven 1½-pounder cannons by 14 horses, four drivers, two 6-pounders, eight horses, four drivers, one kettle drum, two horses, one driver.

Their march was as follows:—

Two ranks of pioneers, six in a rank, with one in the front, one in the centre, and one in the rear.

The tumbrils or covered carts, and the last waggon having a standard on it.

24 mattrosses, under a lieutenant and captain.

Four gunners attending the seven cannon of 1½-pounders.

Two bombardiers attending the two howitzers.

Two gunners attending the four three-pounders.

An adjutant, and one gunner attending the two 6-pounders, the last 6-pounder having a standard on it.

The kettle-drum.

Master artificer and six artificers.

James Deal, captain-lieutenant }
Richard Somerfield, lieutenant } abreast.
John Winch, fireworker }

31 mattrosses in four ranks, six abreast, two in the centre, and one in each corner.

A lieutenant and as many gunners in the same form, with two lieutenants following abreast.

The Earl of Cadogan, general and commander-in-chief of his majesty's forces, master-general of the Ordnance, attended by Colonel Otway as quartermaster-general (in the place of Colonel Armstrong, who, being surveyor-general of the Ordnance, was obliged to attend at the Tower), and Colonel Williamson, adjutant-general.

The troops and artillery having passed the Abby, marched into St. James's, and were drawn up on the parade, where at a signal that the body was deposited, fired three volleys; they then returned to the camp in Hyde Park.

1722. 22nd Sept. Warrant appointing George Harrison, Esq., superintendant of our founderies in our kingdom of Great Britain, and all other our dominions, at a salary of £500 per annum.

9th Nov. A Board's order for a proportion of brass ordnance and stores for a field train for Chester.

Brass ordnance mounted on travelling carriages, with limbers complete ...	3-pounders	2
	1½ "	4
Brass Cohorn mortars on their beds ...	5½-inch	2
	4½ "	6

9th Nov. A field train to be laid up at Edinburgh Castle.

Brass ordnance mounted on travelling carriages, with limbers complete ...	18-pounders	2
	12 "	4
	6 "	2
	4 "	6
	1½ "	6

1722.			
		8 inch howitzers	2
	Brass Cohorn mortars, with their beds ...	5½-inch	4
		4½ do.	10

9 Nov. *A Field Train for Berwick.*

Brass ordnance mounted on travelling carriages with limbers complete ...	3-pounders ...	5
	1½ do.	7
Brass Cohorn mortars, with their beds ...	5½-inch	4
	4½ do.	10

His majesty's warrant for—

1723. 27 May.	The Artillery encamped in Hyde Park with six regiments of Dragoons and twelve regiments of Foot	6-pounders ...	2
		3 do.	4
		1½ do.	14

A Field Train Ordered for Plymouth.

Brass ordnance mounted on travelling carriages with limbers, complete ...	3-pounders of 7ft.	2
	1½ do. 6ft.	4
Brass Cohorns, with their beds	5½-inch	4
	4½ do.	8

A Field Train for Portsmouth.

Brass ordnance mounted on travelling carriages with limbers complete ...	3-pounders ...	4
	1½ do. ...	8
Brass Cohorn mortars with their beds ...	5½-inch	4
	4½ do.	14

1724. Jan. Muster Roll.—Captain Hughes' company at Gibraltar, with the following officers:—

Captain Hughes.
Lieut. Stepken.
,, Anthony Brown.
,, John Winch.

A Field Train Ordered for Hull.

Brass ordnance mounted on travelling carriages with limbers complete ...	3-pounders ...	4
	1½ do. ...	6
Brass Cohorn mortars, with their beds ...	5½-inch	4
	4½ do.	10

Sept. Muster Rolls from Annapolis Royal, the following appear:—

Lieutenant Melledge.
2 bombardiers.
4 gunners.
7 mattrosses.

1724. Extract of a letter from the Honble. Henry Pelham, Esq., his majesty's secretary at war, to Col. Kane, lieutenant-governor at Minorca:—

Sir,

I am now to acquaint you, that representation having been made to his majesty, that the officers of the four regiments in the garrison of Minorca refuse to allow the commission officers of the train of artillery to sit at Courts Martial with them, though the persons to be tryed for offences belong to the train. His majesty taking the same into his consideration is pleased to direct, that as the officers of the trains of artillery are by the 45th Article of War made subject to the said Article, in the same manner as those of the Land Force, It is his royal pleasure, that the commission officers of the train shall sit at Courts Martial, and vote with the commission officers of the army, when any of their own people are to be tryed, and take place according to the rank they bear in the artillery, and to the dates of their commissions, which you are to see regularly complied with.

(Signed) H. Pelham.

1725. John Duke of Argyle and Greenwich appointed master-general

1725. Military branch of the office of Ordnance for year 1725, as relating to the artillery :—

At Gibraltar.

		s.	d.	
1 captain	at	10	0	per day.
1 lieutenant }	„	5	0	„
1 lieutenant }	„	5	0	„
1 lieutenant	„	3	0	„
2 sergeants	„	2	0	each.
2 corporals	„	1	8	„
4 bombardiers	„	1	8	„
14 gunners	„	1	4	„
29 mattrosses	„	1	0	„
2 drums	„	1	0	„

At Mahon.

1 captain	at	10	0	per day.
1 lieutenant	„	6	0	„
1 lieutenant	„	4	0	„
1 fireworker	„	3	0	„
2 sergeants	„	2	0	each.
2 corporals	„	1	8	„
4 bombardiers	„	1	8	„
18 gunners	„	1	4	„
48 mattrosses	„	1	0	„
2 drums	„	1	0	„

At Placentia.

1 lieutenant	at	5	0	per day.
8 gunners	„	1	4	each.

At Annapolis.

		s.	d.	
1 lieutenant	at	5	0	per day.
2 bombardiers	„	1	8	each.
4 gunners	„	1	4	„
7 mattrosses	„	1	0	„

1725. Establishment of the two marching companies in England:—

		£	s.	d.	
1 colonel	at	1	5	0	per day.
2 captains	„	0	10	0	each.
3 first lieutenants	„	0	6	0	„
2 first „	„	0	5	0	„
2 second „	„	0	4	0	„
4 third „	„	0	3	0	„
6 sergeants	„	0	2	0	„
5 corporals	„	0	1	8	„
17 bombardiers	„	0	1	8	„
11 cadet gunners...	„	0	1	4	„
29 gunners	„	0	1	4	„
41 mattrosses	„	0	1	0	„
4 drums	„	0	1	0	„

Master-gunners and gunners belonging to the several garrisons in Great Britain who are warranted by the master-general of the ordnance, and are paid upon the establishment of the Guards and Garrisons:—

		s.	d.	
41 master-gunners	at	2	0	per day each.
178 gunners	„	1	0	each.

Part of the old military establishment in North Britain, still remaining, when it sinks is to be applied to the new military branch:—

		s.	d.	
1 lieutenant	at	5	0	per day.
1 corporal	„	1	3	
4 gunners	„	1	0	each.
1 mattross	„	0	6	
2 bombardiers	„	2	0	„
1 petardier	„	2	0	
2 miners...	„	1	6	„

Allowances annually granted by Parliament to persons to make up their salaries upon the old establishment equal to half the pay they enjoyed in the late wars, &c.:—

		£	s.	d.	
3 colonels	1 at	28	2	6	per annum.
	1 „	1	12	6	„
	1 „	128	2	6	„
1 lieut.-colonel	„	0	7	6	per day.
2 majors...	„	0	7	6	„ each.
1 „	„	36	17	6	per annum

		£	s.	d.	
3 captains	1 at	0	5	0	per day.
	2 „	0	3	0	each.
1 lieutenant	... „	0	1	0	
1 adjutant	... „	0	3	0	
4 fireworkers	... „	0	1	0	each.
1 „	... „	0	2	0	

Part of the old military branch of artillery officers remaining still on the old establishment, which, in proportion as it sinks, is to be applied to the new establishment.

		£	s.	d.	
1 master-gunner of England ...	...	190	0	0	per annum.
1 comptroller of fireworkers ...	...	200	0	0	„
1 firemaster's mate	...	80	0	0	„
2 fireworkers	each	40	0	0	„
2 captains	„	60	0	0	„
1 mate to master-gunner... ...	...	45	10	0	„
1 bombardier	...	36	10	0	„
2 first lieutenants...	each	50	0	0	„
4 gentlemen of ordnance... ...	„	40	0	0	„
1 chief bombardier	...	54	15	0	„
7 bombardiers	each	36	10	0	„
1 chief petardier	...	54	15	0	„
2 petardiers	each	36	10	0	„
33 gunners	„	18	5	0	„

N.B.—This old establishment continued for the year 1726.

A list of superannuated and disabled gunners who have served long and well, being too feeble for active service, are subsisted until they can be placed in the garrisons, &c.:—

		s.	d.	
3 gunners	each at	1	0	per day.
2 bombardiers	„	0	10	„
6 gunners	„	0	8	„
8 mattrosses	„	0	6	„

1726 29 June. Artillery, ordnance and stores embarked on board the squadron under Sir J. Jennings, to assist Gibraltar in case of being besieged.

Two bomb-vessels fitted out to accompany the fleet to the Mediterranean, officers as follows:—

Captain .... Thomas Hughes.
Fireworkers .... Burnett Godfry. / Alexr. Kennedy. / Henry Hopkey.
8 bombardiers.
2 gunners.

1727. 8 Feb. **Gibraltar besieged; 5 March, Capt.-Lieut. Holman killed in one of the batteries. 27th March, a captain, bombardiers,** gunners and mattrosses joined the garrison from England. 21st April, Col. Watson with guns and stores also arrived and took the command of the artillery. The number of gunners and mattrosses at the siege 200.

COL. JAMES'S *History of Gibraltar.*

1727. 15th March. Two bomb-vessels fitted for the Baltick, with the following officers, &c. :—

Lieutenant...............	George Minnens.
Fireworkers	William Backhouse. Thomas Bassett.
1 sergeant.	
1 corporal.	
4 bombardiers.	
4 gunners.	

17th March. Jonas Watson appointed lieut.-colonel in the regiment.

1 April His majesty's warrant to John Duke of Argyle, master-general of the ordnance, to furnish two bomb vessels with mortars, &c., for an expedition by sea.

Brass howitzer mortars	3 for Seaford. 3 „ Shoreham.

	£	s.	d.
1 lieutenant	0	6	0
2 fireworkers, each 3s.	0	6	0
1 sergeant	0	2	0
5 bombardiers, each 1s. 8d.	0	8	4
4 gunners, each 1s. 4d.	0	5	4
2 carpenters, each 2s. 6d....	0	5	0
For the Boats.	per mensem.		
1 coxswain...	2	5	0
4 watermen, each £2 5s. 0d.	9	0	0

20 May His majesty's warrant for a train of artillery of brass ordnance, mortars, &c., with officers, gunners, bombardiers and other attendants, to be sent to Flanders.

Kettle drum waggon, complete	... ...	... 1
Brass cannon	6-pounders ...	... 4
	3 do. ...	.. 12
	1½ do. ...	... 8
Mortars	Royal ...	... 6

List of officers and other ministers appointed to the above :—

	Per diem.		
	£	s.	d.
Colonel	1	5	0
Comptroller	1	0	0
Pay-master...	0	10	0
Adjutant	0	6	0

	Per diem. £	s.	d.
Chaplain	0	6	8
Quarter-master	0	6	0
Commissary of horses	0	6	0
Waggon-master	0	6	0
Surgeon	0	7	6
His mate	0	3	0
Assistant provost-marshal	0	1	4
Kettle drummer	0	3	0
Driver to do.	0	1	4

Company of Artillery.

Captain	0	10	0
Captain-lieutenant	0	7	0
1 lieutenant	0	5	0
2 second lieutenants, each 4s.	0	8	0
3 sergeants, each 2s.	0	6	0
3 corporals, each 1s. 8d.	0	5	0
3 bombardiers, each 1s. 8d.	0	5	0
24 gunners, each 1s. 4d.	1	12	0
48 mattrosses, each 1s.	2	8	0
8 pioneers, each 1s.	0	8	0
2 drummers, each 1s.	0	2	0

Artificers.

1 master-carpenter	0	4	0
3 carpenters, each 2s. 6d....	0	7	6
1 master-wheelwright	0	4	0
2 wheelwrights, each 2s. 6d.	0	5	0
1 master-smith	0	4	0
2 smiths, each 2s. 6d.	0	5	0
1 master-tinman	0	4	0
1 tinman	0	2	0

Civil Officers.

Commissary of stores	0	8	0
His clerk	0	3	0
6 conductors, each 2s. 6d....	0	15	0
1 cooper	0	2	6

Pontoon Men.

Bridge-master	0	5	0
2 corporals, each 1s. 10d.	0	3	8
20 pontoon men, each 1s. 4d.	1	6	8

G. R.

His majesty's warrant.

1727. 17 June. Right trusty and right entirely beloved cousin and councillor. We greet you well. Whereas King Charles the Second and King James the Second did make an establishment of the Office of Ordnance in a book entitled Instructions for the Government of the Office of our Ordnance under our Master-General thereof, committed to five Principal Officers, and signed by the said King Charles and King James and the Principal Secretaries of State, which were confirmed by his majesty King William, her majesty Queen Anne and his late majesty King George, our royal predecessor: Our will and pleasure and express command is, that all orders, warrants, and instructions or directions therein contained and given by the said King Charles and King James aforesaid, shall continue in force, to be obeyed and observed by all and every person or persons who in our said office are and shall be concerned, as fully and amply as the same were or ought to be observed in the reigns of the said King Charles, King James, King William, Queen Anne, and his late majesty King George; and for so doing this shall be your sufficient warrant untill our further pleasure is known. Given at our Court at Leicester House, this 17th day of June 1727, in the first year of our reign.

By His Majesty's command.

HOLLES NEWCASTLE.

To our right trusty and right entirely beloved cousin and councillor John Duke of Argyle and Greenwich, Master-General of our Ordnance.

1 Nov. Colonel Borgarde appointed colonel commandant, the first officer holding that situation.

Establishment of the Royal Artillery :—

1 colonel commandant, Brig. A. Borgarde.
1 lieutenant-colonel, Jonas Watson.
1 major, William Bousfield.
4 companies.
1 adjutant.
1 quarter-master.
1 bridge-master.

Strength of each company :—

1 captain.
1 capt.-lieut.
1 first lieut.
1 second lieut.

3 lieut. fire-wkrs.
3 sergts.
3 corporals.
8 bombrs.
20 gunners.
64 mattrosses.
2 drummers.

1728. Dec. Establishment of the artillery taken from the muster roll of December quarter in England, being two marching companies.

		Per day. £	s.	d.
Colonel	Brigadier A. Borgard	1	5	0
Lieut.-colonel	Jonas Watson	1	0	0
Major	James Petit			
Captains	James Richards	0	10	0
	Thomas Hughes	0	10	0
Captain-lieutenants	Jonathan Lewis	0	6	0
	George Mennins	0	6	0
First lieutenants	Peter Stepkins	0	6	0
	John Forbes	0	5	0
Second lieutenants	George Michelson	0	4	0
	Thomas James	0	4	0
Fireworkers	Henry Hopkey	0	3	0
	John Winch	0	3	0
	George Williamson	0	3	0
	Lyton Leslie	0	3	0
5 sergeants	each	0	2	0
4 corporals	do.	0	1	8
16 bombardiers	do.	0	1	8
16 cadet gunners	do.	0	1	4
32 gunners	do.	0	1	4
50 mattrosses	do.	0	1	0
4 drummers	do.	0	1	0
Surgeon	Francis Lapouge	0	4	0
Mate to do.	Thomas du Pau	0	2	6

At Gibraltar.

		£	s.	d.
Captain	James Deal	0	10	0
Lieutenants	Anthony Brown	0	5	0
	John Washington	0	4	0
	Abraham Taylor	0	3	0
2 sergeants	each	0	2	0
2 corporals	do.	0	1	8
5 bombardiers	do.	0	1	8
19 gunners	do.	0	1	4
88 mattrosses	do.	0	1	0
2 drummers	do.	0	1	0

At Mahone.

		Per diem. £ s. d.
Captain	Thomas Pattison	0 10 0
Lieutenants	George Aikinhead	0 6 0
	Robert Braman	0 4 0
Fireworker	Ivander Mackay	0 3 0
2 sergeants	each	0 2 0
2 corporals	do.	0 1 8
6 bombardiers	do.	0 1 8
20 gunners	do.	0 1 4
51 mattrosses	do.	0 1 0
2 drummers	do.	0 1 0

At Placentia.

Lieutenant	Abraham Kennesbe	0 3 0
8 gunners	each	0 1 4

At Annapolis.

Lieutenant	John Millidge	0 3 0
2 bombardiers	each	0 1 8
4 gunners	do.	0 1 4
8 mattrosses	do.	0 1 0
Chief engineer	John Armstrong	1 7 6
Director	Thomas Lascelles	1 0 0
Engineer in ordinary	John Romer	0 10 0
Sub-engineers	Joseph Day, Thomas Armstrong, William Skinner ... each	0 4 0
Practitioner engineers	Thomas Moore, Robert Bousfield, James Wybault, Bloom Williams, Charles Campbell ... each	0 3 0

Remaining of the Old Establishment.

		Per annum. £ s. d.
Third engineer	Christian Lilly	200 0 0
Engineers	Peter Carles, James Moore, Thomas Phillips, John Stanway, John Selioke, Richard King ... each	100 0 0
	Theodore Dury	77 15 0
Sub-engineer	Benjm. Withall	50 0 0
Master gunner of England	James Pendlebury	190 0 0
Comptroller of fireworks	John Henry Hopkey	200 0 0
Fire-master's mate	John Baxter, sen.	80 0 0

		Per annum. £ s. d.
Fireworkers	Zachariah Smith	40 0 0
	James Finney	40 0 0
	Edmond Williamson	60 0 0
	Wm. Bousfield 60 0 0	
Mate to master gunner	„ „ 45 10 0	142 0 0
Bombardier	„ „ 36 10 0	
First Lieutenant	Ralph Wood	50 0 0
Gentlemen of the ordnance	Rene La Combe; John Palmer; John Aldern; Danl. Coenen	each 40 0 0
Bombardiers	John Kelly; William Haeys; Godfry Franks; John Pawlet; Thomas Marwood; John Green	each 36 10 0
Chief petardier	George Musgrave	54 15 0
19 gunners		each 18 5 0

1729. 8 April. Two bomb vessels fitted out to serve in the Mediterranean.

Lieutenant .... Peter Stepkin.

Fireworkers .... John Milledge. George Williamson. Lytton Leslie.

7 bombardiers.
3 cadet gunners.
5 gunners.
5 mattrosses.

1730. 31 Dec. Stations of the field officers and companies :—

Colonel A. Borgarde, brigadier ...... at Woolwich.
Lieutenant-Colonel Watson, colonel ...... at Woolwich.
Major, William Bousfield .... at Woolwich.
Captains:
James Richards .... at Woolwich.
Thomas Hughes .... at Woolwich.
James Deal .... Gibraltar.
Thomas Pattison .... Minorca.

1732. Feb. The three following gentlemen were preferred by the Duke of Argyle and Greenwich in the Royal Regiment of Artilllery viz. :—

George Williamson, gentleman, to be lieutenant in Captain Pattison's company at Minorca.

John Goodyear, gentleman, to be captain-lieutenant of Captain Deal's company at Gibraltar.

William Sumpter, gentleman, to be lieutenant of Captain Hugh's company at Woolwich.

1733. May. James Cockburne, Esq., paymaster to the Royal Regiment of Artillery, made secretary to the Duke of Argyle as master-general of the ordnance.

1734. Feb. Lieutenant Lewis promoted to captain in Royal Artillery. Henry Hopkey, Esq., made lieutenant of artillery at Port Mahon.

9 Apl. Mr. Felton, storekeeper of the ordnance and Royal Foundry at Woolwich, died. Mr. George Campbell promoted storekeeper of the Royal Foundry, Woolwich.

Nov. Estimate for the artillery:—

	Per diem. £	s.	d.
1 colonel	1	5	0
1 lieut.-colonel	1	0	0
1 major	0	15	0
2 captains, each	0	10	0
2 captain-lieutenants, each	0	6	0
1 first lieutenant	0	6	0
1 do. do.	0	5	0
2 second lieutenants, each	0	4	0
4 lieutenant fireworkers, each	0	3	0
6 sergeants, each	0	2	0
6 corporals, each	0	1	8
16 bombardiers, each	0	1	8
16 cadet gunners, each	0	1	4
36 gunners, each	0	1	4
74 mattrosses, each	0	1	0
4 drummers, each	0	1	0
1 surgeon	0	4	0
1 surgeon's mate	0	2	6

At Mahon.

	£	s.	d.
1 captain	0	10	0
1 first lieutenant	0	6	0
1 second lieutenant	0	4	0
1 lieutenant fireworker	0	3	0
2 sergeants, each	0	2	0
2 corporals, each	0	1	8
6 bombardiers, each	0	1	8
20 gunners, each	0	1	4
51 mattrosses, each	0	1	0
2 drummers, each	0	1	0

At Gibraltar.

	£	s.	d.
1 captain	0	10	0
1 first lieutenant	0	6	0
1 second lieutenant	0	4	0

	Per diem. £	s.	d.
1 lieut. fireworker...	0	3	0
2 sergeants, each ...	0	2	0
2 corporals, each ...	0	1	8
5 bombardiers, each	0	1	8
19 gunners, each ...	0	1	4
38 mattrosses, each	0	1	0
2 drummers, each ...	0	1	0

At Placentia.

	£	s.	d.
1 lieut. fireworker...	0	3	0
1 sergeant ...	0	2	0
7 gunners, each ...	0	1	4

At Annapolis.

	£	s.	d.
1 lieut. fireworker...	0	3	0
1 sergeant ...	0	2	0
1 bombardier	0	1	8
4 gunners, each ...	0	1	4
8 mattrosses, each...	0	1	0

1734. May. Barradale Golesworth, Esq., promoted first commissioner of the Foundry at Woolwich; James Tillie, Esq., as superintendent of the Royal Foundry.

1735. 17 Dec. GEORGE R.

Whereas it hath been most humbly represented unto us, that Lewis Lucas, a private soldier in Captain Thomas Hughes's company, in our Royal Regiment of Artillery, commanded by our trusty and well-beloved Brigadier Albert Borgard, is now in confinement in the Savoy prison for desertion, which we have thought fit should be inquired into by a court martial.

Our will and pleasure therefore is, that a court martial be forthwith held upon this occasion, which is to consist of our trusty and well-beloved Colonel Jonas Watson, lieut.-colonel of our train of artillery, whom we do appoint to be president thereof, of six commissioned officers belonging to our said Royal Regiment of Artillery, and of six commission officers belonging to our three regiments of foot guards who are now in town and can be conveniently summoned to attend the same. And you are to order the provost martial general, or his deputy, to give notice to the said president and officers, and all others whom it may concern, when and where the said court martial is to be held, and to summon such witnesses as shall be able to give testimony in this matter, the said provost martial and his deputy being hereby directed to obey your orders, and give their attendance where it shall be requisite. And we do further authorize and impower the said court martial to hear and examine all such matters and informations as shall be brought before them touching the desertion of the said Lewis Lucas, and proceed in his trial, and in giving sentence according to the rules of military discipline, which said sentence you are to return to our Secretary at War to be laid before us for our consideration, and for so doing this shall be as well to you, as to the said court martial and all others concerned, a sufficient warrant. Given at our Court at St. James's, this 17th day of December, 1735, in the ninth year of our reign.

By His majesty's command,

HARRINGTON.

To our trusty and well-beloved Sir Henry Houghton, Bart., Judge Advocate, General of our Forces, or his Deputy.

1736. March. George Williams, Esq., captain in the royal train of artillery.

4 Dec. Major William Bousfield, Esq., (at Greenwich) of the train of artillery, died. He behaved with great gallantry in King William and Queen Ann's wars, both by sea and land. By sea anno 1694, against Kilmore Castle near Londonderry; anno 1694 at the bombardment of Diepe, Havre de Grace, and at the expedition to Vigo. In Ireland at the battles of the Boyne, Agrim, Limerick, Gallway and Drogheda. At most of the sieges in Flanders, and at the battles of Landen, Hochstet, Ramilies, Oudenarde, &c.

1737. Feb. Promoted.—Lieutenant Simpson, captain of the artillery of Port Mahon.

1738. Three bomb vessels ordered to be in readiness for sea.

Captain..................	Thomas James.
Lieutenants,............	William Belford. John Milledge. John Sutton. Christopher Wilberry. Thomas Mitchell. Lawrence Pitts.

3 sergeants.
9 bombardiers.
3 cadets.
6 gunners.
15 mattrosses.

22 Aug. These vessels sailed from Woolwich in order to join the rest of his majesty's ships of war at Spithead, the place of rendezvous; each bomb vessel carries one 13-inch mortar and one 10-inch howitzer, with 400 shells and 40 carcasses for each mortar and howitzer. The former, with 30 lbs. of powder will throw a shell of

240 lbs. 4,000 yards ; and the latter, with 12 lbs. of powder, a shell of 96 lbs. the same distance.

1738. July. Promoted.—Captain Thomas Bing, captain in the royal artillery. Captain Newton, captain-lieutenant in room of

Aug. Captain Patterson, appointed major of the artillery.

1739. 26 Feb. *Officers of the Royal Regiment of Artillery.*

		Britain.	Gibraltar.	Mahon.
Captains	James Deal	...	1	...
	Jonathan Lewis	1	...	...
	George Michelson	...	...	1
	Thomas James......B	1	...	...
Captain-Lieuts. 6s.	George Minnins, John Milledge...	2	...	...
First Lieuts. 6s.	George Williamson	...	...	1
	Lytton Leslie ...	...	1	...
First Lieuts. 5s.	John Forbes ...	1	...	...
	Ivander Mackey	...	...	1
Second Lieuts. 4s.	William Belford	1	...	...
	Borgarde Michelson ...	...	...	1
	William Sumpter ...	7	...	...
	John Goodyer......B ...			
	Christopher Wilbury ...			
	Withers Borgarde ...			
	Archibald Cunningham (B)			
	Thomas Mitchell ... (B)			
	Lawrence Pitts ... (B)			
	James Mace ...	...	1	...
	John Sutton ...	1	...	...
	Charles Lodwick	...	1	...

Wanting to complete the establishment at home, Gibraltar, &c.

2 lieutenants at 4s.

2 lieutenant-fireworkers at 3s.

1 lieutenant-fireworker ... at Annapolis

1 " " ... at Placentia

July. Major-General Albert Borgard promoted to lieutenant-general.

Colonel Armstrong promoted to major-general.

Thomas Sydenham, a clerk to the lieutenant-general of the ordnance, at £150 per annum.

6 July. Francis de la Rochefoucaut, Marquis de Montandre, died, a field marshal, general of the ordnance in Ireland, and governor of Guernsey.

1739. August. Major-General Armstrong promoted master of the ordnance in Ireland, in room of the Marquis Montandre.

5 Sept. Three bomb vessels fitted for sea, two of them for the West Indias, the third for the Mediterranean.

Captain..................	Thomas James.
Lieutenants............	John Sutton. Thomas Mitchell. Lawrence Pitts. John Goodyer.

2 sergeants.
6 bombardiers.
1 cadet.
5 gunners.
6 mattrosses.
2 carpenters.
5 watermen.

List of the artillery :—

Two Companies at Home.

1 colonel	Major-General A. Borgard.
1 lieutenant-colonel	Jonas Watson.
1 major...........................	Thomas Pattison.
2 captains	Jonathan Lewis. Peter Stepkins.
2 captain-lieutenants	George Minnens. Thomas James.

1 first lieutenant.
1 „
2 second lieutenants.
4 lieutenant-fireworkers.
6 sergeants.
6 corporals.
16 bombardiers.
16 cadet gunners.
40 gunners.
92 mattrosses.
4 drummers.
1 surgeon.
1 surgeon's mate.

At Mahon.

1 captain	George Michelson.
1 lieutenant	George Williamson.
1 „	Ivander Mackay.
1 lieutenant-fireworker	Borgard Michelson.

2 sergeants.

2 corporals.
6 bombardiers.
20 gunners.
51 mattrosses.
2 drummers.

At Gibraltar.

1 captain .... James Deal.
1 lieutenant..................... Lytton Lesslie.
1 „ .... Archibald Cunningham.
1 lieutenant-fireworker ...... James Mace.
2 sergeants.
2 corporals.
5 bombardiers.
19 gunners.
38 mattrosses.
2 drummers.

At Placentia.

1 lieutenant-fireworker.
1 sergeant.
8 gunners.

At Annapolis.

1 lieutenant.
1 sergeant.
1 bombardier.
4 gunners.
8 mattrosses.

Names of the officers of the two companies at Woolwich:—

Captain	Jonathan Lewis	at Woolwich.
Captain-lieutenant ...	John Melledge	lame, and has been so these six months.
First lieutenant	John Forbes	lost his memory, done no duty these seven years.
Second lieutenants......	William Belford	gone recruiting.
	John Goodyer	on board the bomb vessels.
Lieutenant-fireworkers	Withem Borgard.........	at Woolwich.
	Lawrance Pitt............	on board the bomb vessels.
Sergeants	2	invalids at Woolwich.
	1	on board the bombs.
Corporals.................	2	at Woolwich.
	1	recruiting.
Bombardiers	3	on board the bombs.
	5	at Woolwich.
Cadet gunners	Richard Temple	a very idle fellow.
	Charles Otway............	I know not where he is.
	Peter Bruce...............	at Woolwich.
	Osborn Sidney Wandesford	I know not where they are.
	Henry Tichbourn	
	John Richards............	on board the bombs.
	Archibald Douglass ...	I know not where they are.
	James Chamberlaine ...	

Gunners	15	at Woolwich, 5 to be made bombardiers on augmentation.
	3	invalids.
	2	on board bombs.
Mattrosses	41	at Woolwich, 25 to be made gunners at augmentation.
	1	invalid.
	1	at Whitehall, invalid.
	3	on board bombs.
Drummers	2	at Woolwich.

Captain	Thomas James	on board the bombs.
Captain-lieutenant	George Minnins	old and worn out in the service.
Second lieutenant	William Sumpter	at Woolwich.
Lieutenant-fireworkers	Charles Welbury	at Woolwich.
	Thomas Mitchell	on board bombs.
	John Sutton	on board bombs.
Sergeants	2	at Woolwich.
	1	on board bombs.
Corporals	1	gone to fetch recruits from Nottingham.
	1	recruiting.
	1	at Woolwich.
Bombardiers	1	at Woolwich, invalid.
	3	in bomb vessels.
	4	at Woolwich.
Cadet gunners	Moses Baxter	at Portsmouth.
	John Laye	at Portsmouth.
	John Friske	I know not where they are.
	William Green	I know not where they are.
	John Chalmers	at Woolwich.
	Leonard Smelt	in the Tower.
	Sir Robert Richardson	I know not where they are.
	Charles Soley	I know not where they are.
Gunners	16	at Woolwich, 5 to be bombardiers at augmentation.
	1	at Woolwich, invalid.
	3	on board bombs.
Mattrosses	44	at Woolwich, 25 to be gunners at augmentation.
	1	at Woolwich, invalid.
	2	at Whitehall, invalid.
	3	on board bomb vessels.
Drummers	1	recruiting.
	1	at Woolwich.

1739. 14 Sept. His majesty's warrant for an augmentation to the two companies at home.

An account shewing the number of officers and men of which the two marching companies of the Royal Regiment of Artillery, on the *home* establishment, consist, with a proposed augmentation thereof.

1 colonel .... A. Borgard.
1 lieutenant-colonel .... Jonas Watson.
1 major............................... Thomas Pattison.
2 captains.
2 captain-lieutenants.
8 lieutenants and fireworkers.
6 sergeants.
6 corporals.
16 bombardiers.
16 cadet gunners.
40 gunners.
92 mattrosses.
4 drummers.
1 surgeon .... Peter Laponge.
1 surgeon's mate.................. Thomas Depaue.

Augmentation proposed, with their pay per diem :—

	£	s.	d.
1 bridge-master .... John Chalmers	0	5	0
2 corporals..each	0	2	6
2 tinmen .... „	0	2	6
20 pontoon men .... „	0	1	4
10 bombardiers .... „	0	1	8
20 gunners... „	0	1	4
24 mattrosses.. „	0	1	0

1740.
1 Jan.

Companies at

Minorca.	Gibraltar.
Captain ...... G. Michelson.	James Deale.
Lieutenants { G. Williamson. / Ivander Mackey.	Lieutenant ...Lytton Leslie.
2nd lieut. ... Borgd. Michelson.	Lieutenant fireworkers { Archd. Cunningham. / James Mace. / Ch. Lodwick.
2 sergeants.	1 sergeant.
2 corporals.	2 corporals.
5 bombardiers.	5 bombardiers.
1 „	
19 gunners.	17 gunners.
51 mattrosses.	39 mattrosses.
2 drummers.	2 drummers.
1 mattross.	

1740. 1 Jan.

Captain	Thomas James	on board bomb vessels.
Lieutenant	J. Goodyear	
Lieutenant fireworkers ...	Thomas Mitchell	
	Law. Pitts	
	John Sutton	
2 sergeants		
6 bombardiers		
1 cadet gunner ...	John Richards	
5 gunners		
6 mattrosses..		
1 carpenter		

1 April. Pay of officers and three companies at Woolwich:—

Colonel...........................	A. Borgarde.
Lieutenant-colonel	J. Watson.
Major	Thomas Pattison.
Captains	J. Lewis.
	J. Milledge.
Captain-lieutenant	G. Minnins, commanding Capt. James's company.
First lieutenants	William Belford.
	John Forbes.
	William Sumpter.
	Christopher Welbury.
Second lieutenants	William Borgard.
	G. A. Elliot.

John Chalmers.
Moses Baxter.
Henry Watson Coliere.
Thomas Flight.
James Pattison.
Thomas Desaguliers.
Samuel Shepherdson.
Thomas Ord.
David Rogers.
John Godwin.

1 July. Company with miners:—

Captain	J. Milledge.
First lieutenants	W. Belford.
	Christopher Welbury.
Second lieutenant	Isaac Delanawze.
Lieutenant fireworkers ...	David Rogers.
	John Godwin.
	Thomas Ord.

4 sergeants.	70 mattrosses.
4 corporals.	2 drummers.
11 bombardiers.	21 miners.
88 gunners.	

1740. 31 Jan. His majesty's warrant for delivering to Colonel William Blackeney, ordered on a special commission, three thousand stand of arms, ammunition, &c., to be put on board such ship at Portsmouth as he may appoint.

29 May. His majesty's warrant for furnishing Captain George Anson with brass ordnance, mortars, &c., for a particular expedition.

Field pieces, complete	3-pounders	4
Coehorns	$4\frac{2}{5}$-inch	20
To attend this service	3 gunners. 3 bombardiers.	

29 May. His majesty's warrant for ordnance and stores to the three camps near Hounslow, Newberry and Windsor.

Proportions for

Hounslow.	*Windsor.*	*Newberry.*
Lieut. .... Pattison.	Lieut....Thomas Flight.	Lieut. ...Desaguliers.
1 conductor of stores.	with the same proportion as for Hounslow.	
1 sergeant.		
3 corporals or bombardiers.		
6 gunners.		
12 mattrosses.		
1 drummer.		
1 wheelwright.		
1 conductor of horses.		
13 carters.		

			Horses for the same.
Brass cannon mounted on travelling carriages (without limbers) with 1 bed, 2 coines, 1 muzzle tampeon, 1 ladle, 1 rammer, 1 spunge, 1 wadhook, and 1 apron of lead to each.	6-pounders	2	6
	3 „	3	9
Spare carriages and ammunition waggons			7
Waggons			6
Tumbrills			4
Forge carts			3

Lieutenant Chambers appointed (in January's Gazette) as master of the pontoons to the marines.

Feb. John Chalmers appointed (by Gazette) bridge master to the Royal Regiment of Artillery.

1740. 1 July. Captain John Milledge's company and the miners ordered on an expedition to the West Indias.

Oct. An expedition under Sir C. Ogle and Lord Cathcart, in which were Colonel Watson, Major Lewis, and four companies of artillery, sailed from Portsmouth to reinforce Admiral Vernon in the West Indias.

19 Dec. Lord Cathcart died on reaching Dominica, and was succeeded by General Wentworth.

22 May. Jonas Watson to be colonel of the train of artillery ordered on an expedition.

Jonathan Lewis to be major of the same.

John Duke of Montague appointed master-general.

21 Apl. His majesty's warrant directing that the companies of artillery at Mahon and Gibraltar be augmented to 100 men each, and the two companies at home to 150 men each, with an addition of officers and fireworkers, and a third company to be raised to consist of 150 men besides officers.

The augmentation to each company was as follows :—

At Home.	*At Gibraltar.*	*At Mahon.*
1 second lieutenant.	1 first lieutenant.	1 first lieutenant.
2 fireworkers.	2 fireworkers.	2 fireworkers.
1 sergeant.	1 sergeant.	1 sergeant.
9 bombardiers.	1 corporal.	1 corporal.
10 miners.	7 bombardiers.	6 bombardiers.
8 mattrosses.	7 gunners.	6 gunners.
	16 mattrosses.	3 mattrosses.

The New Company.

1 captain.	38 gunners.
1 first lieutenant.	10 pontoon men.
1 „	61 mattrosses.
2 second lieutenants.	2 drummers.
4 fireworkers.	1 tinman.
4 sergeants.	10 miners.
4 corporals.	
20 bombardiers.	

14 March. His majesty's warrant for a train of artillery to attend the land forces on an expedition with Lord Cathcart :—

Brass ordnance mounted on travelling carriages complete...	24-pounders	... 8
	12 do.	... 10
	6 do.	... 4
	3 do.	... 8

Brass mortars	10-inch	2
	8 do.	2
Coehorns (not sent)	Large	10
	Small	40
Petards...	Small	2

A list of the officers and ministers and other attendants appointed to attend the train:—

	Per diem. £	s.	d.
Colonel, Jonas Watson, appointed 22nd May	1	5	0
Major, Jonathan Lewis, appointed 10th July	0	15	0
Adjutant	0	6	0
12 engineers, each 10s.	6	0	0
Pay-master	0	8	0
Commissary of stores	0	8	0
His clerk	0	4	0
6 conductors, each 2s. 6d.	0	15	0
Surgeon	0	5	0
His mate	0	3	0

A Company of Gunners consisting of

	£	s.	d.
Captain	0	10	0
First lieutenant	0	6	0
2 second lieutenants, each 5s.	0	10	0
4 sergeants, each 2s. 6d.	0	10	0
4 corporals, each 2s.	0	8	0
30 gunners, each 1s. 6d.	2	5	0
60 mattrosses, each 1s.	3	0	0
4 fireworkers, each 4s.	0	16	0
8 bombardiers, each 2s.	0	16	0
2 drummers, each 1s.	0	2	0
Master-carpenter	0	4	0
4 carpenters, each 2s. 6d.	0	10	0
Master-wheelwright	0	4	0
2 wheelwrights, each 2s. 6d.	0	5	0
2 coopers, each 3s.	0	6	0
2 farriers, or smiths, each 2s. 6d.	0	5	0
2 collar-makers, each 2s. 6d.	0	5	0

A Company of Miners.

	£	s.	d.
Lieutenant...	0	4	0
20 miners, each 1s. 6d.	1	10	0
Commissary of horses and pay-master's assistant ...	0	6	0
Assistant to the commissary	0	3	0

A chief engineer appointed by warrant for this expedition the 18th April, at 20s. per diem.

1740. 1 July A warrant for an addition of mortars and ordnance stores to the proportion of the train of artillery on the expedition with Lord Cathcart.

Coehorn mortars	$5\frac{1}{2}$-inch	10
	$4\frac{2}{5}$ do.	10

2 Sept. Warrant for a further addition of stores for the service of the expedition with Lord Cathcart.

For the use of Colonel Watson and Captain Moore.

Field bedsteads with curtains, compleat		2
Mattrasses		2
Bolsters and pillows	each	1
Blankets...	pair	1
Quilt		1
Carpets		2
Iron chest for the commissary		1
6 oar'd boat, compleat for the engineer		1

With other laboratory and collar-maker's stores.

July. Lord Mark Kerr appointed general of the ordnance in Ireland.

1741. 31 Oct. The Royal Military Academy was established by the Duke of Montague, then master-general of the ordnance, for instructing the gentlemen of the train of artillery.

During the campaigns in Flanders the great kettle drums, mounted on a triumphal car, finely ornamented and drawn by six horses, marched at the head of the train of artillery. In these campaigns the artillery of all nations forming the army marched in one continued column, but whenever the heavy baggage was ordered away from the army, the kettle drums were considered as part of it.

25 Jan. West Indias. Admiral Vernon and the expedition under General Wentworth sail from Jamaica.

4 March. The expedition anchored off Carthagena.

10 March. The forces were landed and began an attack on the forts. Artillery and stores landing and batteries preparing.

22 March. Opened a fascine battery of twenty 24-pounders against the Castle of Bona Chica.

25 March. The castle carried by storm.

26 March. The army embarked on board the men of war to proceed up the lake, to attack a strong castle, three forts, a boom, four ships of the line, and two batteries, the whole amounting 370 pieces of

cannon. In this attack Colonel Watson, commanding the artillery, was killed, with several other officers of the army and 400 men.

1741. 5 April. The army landed to attack the forts, and became extremely sickly.

9 April. A general attack on the enemies' forts (without opening trenches or making a breach), which were so well defended and such an incessant discharge of grape shot from the walls, that the general was obliged to order a retreat.

The mortality in the camp swept off such numbers that it was thought necessary to raise the siege.

26 April. The demolition of the forts and castles first captured being completed, the troops were re-embarked.

Major Lewis, on succeeding to the command of the artillery (on the death of Colonel Watson), appointed himself a lieutenant-colonel until the pleasure of his majesty was known; he offered his senior captain a majority, who declined, as did the other three: he then gave it to Captain-Lieutenant Belford; both commissions were confirmed at home.

30 April. His majesty's warrant dividing the three marching companies at home into four.

Proposed establishment of the four marching companies:—

		£	s.	d.
1 colonel		1	5	0
1 lieutenant-colonel		1	0	0
1 major		0	15	0
1 adjutant and bridge-master		0	5	0
1 chaplain		0	6	8
1 surgeon		0	4	0
1 surgeon's mate		0	2	6
4 tinmen	each	0	2	6
4 captains	„	0	10	0
4 captain-lieutenants	„	0	6	0
4 first lieutenants	„	0	5	0
4 second lieutenants	„	0	4	0
12 lieutenant-fireworkers	„	0	3	0
12 sergeants	„	0	2	0
12 corporals	„	0	1	8
56 bombardiers	„	0	1	8
20 miners	„	0	1	8
92 gunners 16 cadet gunners	„ „	0	1	4
32 pontoon men	„	0	1	4
184 mattrosses 16 cadet mattrosses	„ „	0	1	0
8 drummers	„	0	1	0

1741. 18 June. His majesty's warrant for a train of artillery to be sent to the camp at Lexden Heath near Colchester.

Ordnance	6-pounders	2
	3-pounders	4

1 lieutenant-fireworker.
1 sergeant.
1 corporal.
2 bombardiers.
6 gunners.
12 mattrosses.
1 drummer.

1742. Feb. Lord Visc. Cobham appointed lieutenant-general of the Ordnance.

Feb. John Duke of Argyle and Greenwich appointed master-general; he resigned the following month.

March. John Duke of Montague succeeded him.

7 March. His majesty's warrant, adverting to the warrant of the 30 April, 1741, directing the three marching companies to be divided into four, and the master-general's further representation that the company established by the warrant 14 March, 1739, to attend the expedition in the West Indias, and lately returned, should be incorporated into the regiment, and that those five companies should be divided into six companies according to the list as follows.

The above six companies, with the companies at Mahon and Gibraltar, form the establishment, with :—

1 colonel.
1 lieut.-colonel.
1 major.
1 quarter-master.
1 adjutant.
1 bridge-master.
1 chaplain.
1 surgeon.
1 surgeon's mate.

15 April. Died.—John Armstrong, Esq., major-general of the forces, colonel of the royal regiment of foot in Ireland, surveyor-general of the Ordnance, and his majesty's chief engineer.

Thomas Lascelles, Esq., director of engineers, appointed master-surveyor of the ordnance, in room of General Armstrong, deceased.

June. Thomas Lascelles, Esq., appointed chief engineer of Great Britain, in room of General Armstrong, deceased.

General George Wade appointed lieutenant-general of the Ordnance.

1742. 18 June. A detachment of artillery embarked at Woolwich for Flanders with 24 heavy 3-pounders.

30 June. They arrived at Ghent.

Aug. A frigate, with a detachment of artillery and five hundred men of General Wentworth's army, took possession of the Island of Rattan in the Bay of Honduras.

1743.[1] June. The Royal Regiment of Artillery consisted of eight companies, commanded by

1 colonel (Lieut.-General Albert Borgard).
1 lieut.-colonel (Thomas Pattison).
1 major (Jonathan Lewis).
1 chaplain.
1 adjutant.
1 quartermaster.
1 bridge master.
1 surgeon.
1 surgeon's mate.[2]

The establishment of each company was

1 captain.
1 captain-lieutenant.
1 first lieutenant.
1 second lieutenant.
3 lieutenant fireworkers.
3 sergeants.
3 corporals.
8 bombadiers.
20 gunners.
64 mattrosses.[3]
2 drummers.

making 107 of all ranks in each company.

The uniform dress of the officers was a plain blue coat, lined with scarlet; a large scarlet Argyle cuff, double-breasted, and with yellow buttons to the bottom of the skirts, scarlet waistcoat and breeches, the waistcoat trimmed with a broad gold lace, and a gold laced hat.

The sergeants coats were trimmed,—the lappels, cuffs, and pockets with a broad single gold lace; the corporals and bombardiers with a

---

[1] Pages 228 to 242 inclusive, are reprinted from the Occasional Papers, R. A. Institution, vol. 1, p. 380.

[2] The first who had a commission of Colonel of the Royal Regiment of Artillery, was Lieut.-General Albert Borgard; it is dated Nov. 1, 1727, and the regiment then consisted of four companies.

[3] The matross assisted in spunging and loading, but his duty was rather to protect than to serve the gun.

narrow single gold lace; the gunners and matrosses, plain blue coats; all the non-commissioned officers and privates having scarlet half-lappels, scarlet cuffs, and slashed sleeves with five buttons, and blue waistcoats and breeches; the sergeants hats trimmed with a broad, and the other non-commissioned officers and privates, with a narrow gold lace. White spatterdashes were then worn. The regimental clothing was delivered to the non-commissioned officers and privates once a year, excepting regimental coats, which they received only every second year; and the intermediate year a coarse blue surtout which served for laboratory works, cookings, fatigues, &c., was delivered with the usual small mounting.

The arms of the officers were fuzees without bayonets, and not uniform. The sergeants, corporals, and bombardiers, were armed with halberts, and long brass-hilted swords; the gunners carried field staffs about two feet longer than a halbert, with two linstock locks branching out at the head, and a spear projecting between and beyond them. Great attention was paid in keeping these very bright. A buff belt over the left shoulder, slinging a large powder horn, mounted with brass over the right pocket, and the same long brass-hilted swords worn by the non-commissioned officers. The matrosses had only common muskets, with bayonets and cartouche-boxes.

The regiment was distributed in the following manner, viz:—one company in Minorca, one at Gibraltar, one at the fishing posts in Newfoundland, three with the army in Flanders, and two at Woolwich;—the two last furnished all detachments, which were only those on board the bomb vessels on service with the fleets in the Mediterranean and West Indies.

1744. March. Two companies were raised and added to the regiment, which now consisted of ten companies, and were disposed as follows, viz:—In Minorca, Gibraltar, and Newfoundland, one company at each; one company was sent to Flanders, in addition to the three already there, and three remained at Woolwich.

June. H.R.H. William, Duke of Cumberland, came to see some experiments, and a proof of guns at Woolwich, when the three companies there were under arms, commanded by Major Lewis. About twenty cadets then at Woolwich were formed in a rank entire on the right of the companies, without arms or uniform, and no officer at their head.

The officers did not then practise saluting with the fuzee, and only pulled off their hats as the Duke passed them.

A circumstance which does so much honour to the corps of Artillery as the following, ought to be mentioned:—

The King of Sardinia, then in alliance with Britain, being threatened with an invasion by the combined armies of France and Spain, applied to Admiral Matthews, commanding the fleet in the Mediterranean, and prevailed with him to consent that the detachments of artillery serving on board the four bomb-ketches in the fleet, should be sent on shore to take charge of the most important posts and batteries on his frontier, from a persuasion, as may be justly presumed, that they were more to be depended on than his own artillerists.

This detachment, consisting of one captain, four lieutenants, and twenty-four bombardiers, performed all that was required of them, perfectly to the satisfaction of his Sardinian Majesty. They were taken prisoners in the defence of Montalban and Montleuze, two strong posts which were taken by the French and Spaniards by assault in April.

Although the Royal Military Academy was established in 1741, by the Duke of Montague, then Master-General, a COMPANY OF GENTLEMEN CADETS was formed and added to the regiment in January, 1745. The cadets mentioned before were, two cadet-gunners, and two cadet matrosses, hitherto mustered with each company; the former received 1s. 4d. and the latter 1s. per diem, paid them monthly by the captains to whose companies they were mustered. A few of these young gentlemen, who were generally the sons of officers, resided at Woolwich, and attended the Royal Academy when they pleased, were under no command, wore no uniform, and were generally so young that few of them were fit to be preferred to commissions:—so little were these young gentlemen under any kind of order, that it was the business of the officer on duty in the warren, who visited the Academy occasionally, to endeavour to preserve good order.

The disposition of the three companies at the fixed stations of Minorca, Gibraltar, and Newfoundland continued the same.

Three companies remained at Woolwich until August, when upon the taking Cape Breton, one of these was sent to garrison; and in the month of October, the four companies in Flanders were ordered home in consequence of the rebellion in Scotland, and were employed in North and South Britain, with the several corps of troops which assembled on that occasion,

A detachment of three officers and fifty men, was sent from Woolwich in August, with a battalion of guards, and the 15th regiment, to strengthen the garrison of Ostend, then besieged by Count Lowendahl, which held out about fourteen days.

A review so different from the present fashion, should not be omitted to be noticed. At the camp of the allied army near Brussels, on receiving the news of the taking of Cape Breton or (Louisburg), the army was drawn up in order of battle, and reviewed by the Duke of Cumberland; the park of artillery was formed in great order, on a fine extensive plain; the four companies of artillery under arms drawn up, two on the right and two on the left of the park. Col. Patterson (then Lieut.-Colonel), Colonel Lewis (then major of the regiment), and Major Belford, then a captain, but doing major's duty to the artillery in Flanders, posted themselves on horseback in front of the park, where they saluted His Royal Highness, as he passed, by dropping their swords; the other officers carrying fusees, only took off their hats as he passed them.

1744. 8 Dec. Two companies with a train of artillery, commanded by Colonel Lewis, marched from Woolwich to Finchley Common where a corps of troops was to have assembled under the immediate orders of the King, had the rebels advanced to London as was then apprehended, but they retired northward from Derby, and these two companies and the train of artillery returned on the 11th to Woolwich.

In this year the British troops in Flanders consisted of twenty-seven battalions and twenty-six squadrons, ten heavy 6-prs., twenty-seven heavy 3-prs., six 1½-prs., and four 8-inch howitzers.

Establishment of the companies (the cadet company having been added this year with a 1st lieutenant, a 2nd lieutenant, and a lieut.-fireworker for its officers), viz.

2 colonels-in-chief and *en seconde*.
3 field officers.
21 captains.[1]
53 subalterns.
1 chaplain.
1 adjutant.
1 quarter-master
1 bridge-master.
2 medical officers.
48 gentlemen cadets.
200 gunners.
640 matrosses.
20 drummers.

Distribution of the 11 companies:—

7 in Great Britain.
1 „ Gibraltar.
1 „ Minorca.
1 „ Newfoundland.
1 „ Louisburg.
11 Total.

1 At this period the cadet company had no captain and only one fireworker instead of three, as with the other companies.

The Master-General of the Ordnance shall have the same respect from the troops with generals of horse and foot, *i.e.* upon all occasions to have the march beat to him, and is to be saluted by all officers, colonels excepted.

The distribution of four companies in Minorca, Gibraltar, Newfoundland, and Louisburg continued the same as last year; one company was stationed in Scotland, and five companies remained at Woolwich, until the suppression of the rebellion in April, by the victory of Culloden, when two of these companies were sent in June with seven battalions to join the allied army in Brabant, and one company embarked in May with six battalions, commanded by Lieut.-General St. Clair, on an expedition to the coast of France, where a fruitless attempt was made on Port l'Orient. This expedition was originally intended against Quebec. Two companies remained then at Woolwich.

N.B. The company detached on the expedition to the coast of France was commanded by Captain Chalmers. He and the company under his command were put, by express orders of the Master-General, under the command of Mr. Armstrong, the chief engineer on that service, who had not at that time, nor even had before or since any rank or commission in the army; for the corps of engineers had no military rank or title until the year 1757; and the circumstance just alluded to is the more remarkable, as hitherto on all services whenever engineers were employed were immediately under the command of the officer commanding the artillery, and always encamped and messed with that corps, and were detached by him as occasions required.

During the campaign in Brabant, two field pieces being heavy 3-prs. commanded by a lieutenant and a detachment of artillery were, for the first time on service attached to each regiment, and encamped with it.

1746. There were no more in Flanders than seven battalions, and nine squadrons, with fourteen heavy 3-prs. for the battalions.

Distribution :—

1	Company,	Gibraltar.
1	"	Minorca.
1	"	Newfoundland.
2	"	Continent.
1	"	Louisburg.
5	"	Great Britain.
11	including cadets.	

Early in 1744, three companies were raised and added to the

regiment, now consisting of thirteen, exclusive of cadets. About the same time three regiments of cavalry being reduced to dragoons (now the three regiments of dragoon guards), and the troopers having it in their option to remain as dragoons or be discharged, many of them chose the latter, and above 200 of them enlisted in the artillery; from this period the regiment improved much in appearance and in the size of the men, neither of which had been much attended to, but receiving at once so many tall men in the corps, may be said to have given rise to the change that has taken place in regard to the height, figure, and strength of the men which now compose it.

The thirteen companies were disposed of during this year as follows: five companies continued stationary at Minorca, Gibraltar, Newfoundland, Louisburg, and in Scotland. The two companies which were sent last year to serve in the army in Brabant, were augmented to five by the junction of two companies from Woolwich, and also of that which was employed last year on an expedition to the coast of France.

One company was embarked on an expedition to the East Indies, commanded by Admiral Boscawen, who failed in an attempt at Pondicherry.

Two companies remained at Woolwich, from which a detachment commanded by a captain-lieutenant, with three lieutenants, and seventy non-commissioned officers and men, were sent in July to assist in the defence of Bergen-op-zoom. This detachment lost fourteen men during the siege.

This year the corps of artillery in Flanders, commanded by Lieut.-Colonel Belford, was well trained in the use of small arms; for the first time there were field days, and the corps had improved so much in the manual exercise and common manœuvres of the infantry, which had never been hitherto practised in the regiment, that his Royal Highness the Duke of Cumberland reviewed the five companies. Upon this occasion the gunners of each company with their field-staffs formed upon the right as a company of grenadiers, and the matrosses with their muskets as a battalion, in which manner they passed in review.

During this campaign, a quarter-guard of the regiment of artillery commanded by a subaltern (as usual in the infantry), was first mounted and encamped opposite the flag-gun,[1] and kettle-drum. A detachment

[1] The "flag-gun" carried the Royal Standard, and indicated the head quarters of the army in the field.

from the infantry, commanded by a captain, with from 60 to 100 men, were always mounted upon the park of artillery, and encamped opposite the centre of the park, being relieved every 48 hours.

Establishment, 14 companies (including cadets)—

2 colonels-in-chief and *en seconde*.
3 field officers.
27 captains.
68 subalterns.
1 chaplain.
1 adjutant.
1 quarter-master.
1 bridge-master.
2 medical officers.
48 gentlemen cadets.
182 non-commissioned officers.
260 gunners.
832 matrosses.
26 drummers and fifers.

1,454

Distribution :—

1 Gibraltar.
1 Minorca.
1 Newfoundland.
5 Continent (Brabant).
1 Louisburg.
1 East Indies (Pondicherry).
1 Scotland.
2 Woolwich.
1 Cadets.

1748. 28 Jan. On Lieut.-Colonel Thomas Patterson, and Major Jonathan Lewis, retiring from the regiment, on account of their old age and infirmities (who were succeeded by Lieut.-Colonel Belford, and Major Borgard Mitchelson), the King signed a warrant authorizing the Master-General to make a provision of full pay for them. The warrant begins thus: "Whereas, the four field officers of the Royal Regiment of Artillery, are the Master-General of our Ordnance, Colonel Commander-in-Chief of the said regiment; a Colonel of Artillery, Colonel-Commandant of the same; a Lieut.-Colonel; and a Major," &c.[1]

[1] Kane's List gives the date of William Belford's commission as Lieut.-Colonel to be February 16, 1749, being one year later than as above. He became Colonel-Commandant of the regiment in March 8, 1751, and died at Woolwich, July 1, 1780. Service 54 years.

1748. 11 Feb. The King signed a warrant directing that from henceforth when any vacancy or vacancies of a matross or matrosses shall happen in any of the companies of the Royal Regiment of Artillery, the said vacancy or vacancies shall be continued until the money arising therefrom shall be sufficient to procure a proper recruit or recruits, and no other allowance shall be made for that purpose.

Kane's list for this year gives the strength per company, the same as in 1745, but there were only sixty-two matrosses instead of sixty-four, and two non-effectives per company, as an allowance to superannuated officers, and for two tinmen.

Before the army took the field this year the gunners' field staves, powder horns with slings, and swords, and the matrosses' muskets, were laid aside, and both these ranks were armed with carbines and bayonets; all the non-commissioned officers still continued to have halberts.

The same distribution and establishment of the companies last mentioned continued, without any material variation, till the latter end of the year, when, upon the peace of Aix-la-Chapelle, and the return of the British troops from Flanders, three entire companies, officers included, were reduced; the officers afterwards came into their former rank in the regiment as vacancies happened—the junior of these came into full pay in 1753.

During the war from 1742 to 1745, Flanders was the great school for military discipline, particularly after the Duke of Cumberland was appointed in 1745, captain-general and commander-in-chief of the British troops, and took command of the allied armies. The strength was as follows:—

	Batt.	Squad.	Heavy 3 prs.			
1742	19	10	24			
1743	—	do.	do.	8-in.	bow.	heavy 6-pr.
1744	22	29	30	4	10	6 1½-prs.
1745	27	26	27	4	10	1 1½-prs.
1746	7	9	14 (Batt. guns.)			
1747	14	14	14	2	14	6 heavy 12-prs.
	with 6 heavy 9-prs.		12 light 6-prs.	6 royal mortars.		
1748	22	14		2	14	6 heavy 12-prs.
66 light 6-prs. for battalions.				6 heavy 9 prs.		6 royal mortars.

This year there were five companies of 100 men,—seven officers each and two field officers.

The foregoing abstract of the troops and artillery in Flanders, during the war of 1741, is given to mark the utility the artillery was found to be of, and the quick progress it made in the addition of it, in proportion to the number of troops employed during that period.

It is justice due to the regiment of artillery to mention that on all occasions during this war, whenever the corps or any part of it was brought to act, their behaviour was such as to distinguish them greatly, —gained them great applause from the troops with whom they served, and thereby established the reputation of the regiment.

The regiment of artillery owes much to the memory of Colonel Belford and Major Borgard Mitchelson, for their zeal and diligence, and their influence during the campaigns in 1747 and 1748, when they commanded in Flanders, for the corps then began to emerge from that state of oblivion and obscurity in which it had hitherto remained. It now began to bear a regular military appearance; great attention was paid to good order, strict discipline and subordination, a change that was far from agreeable to the older officers, who, being promoted from the ranks, had grown up with erroneous notions and bad habits, inconsistent with any military system; but the junior officers, who of late had been promoted from the cadet company, being of a different stamp and better educated, and being now the majority, entered with great zeal and military spirit into the newly adopted alterations and improvements that were introduced by these two officers.

The first fifers in the British service were established in the Royal Regiment of Artillery at the end of this war, being taught by John Ulrich, a Hanoverian fifer, brought from Flanders, by Colonel Belford, when the allied army separated.

Establishment of a company:—

1 captain.
1 captain-lieutenant.[1]
1 first lieutenant.
1 second lieutenant.
2 lieutenant-fireworkers.
3 serjeants.
3 corporals.
8 bombardiers.
20 gunners.
62 matrosses.
2 drummers.
2 non-effectives, for an allowance to superannuated officers, and for 2 tinmen.

Peace.—Three companies reduced:—

---

[1] The senior lieutenant is now called captain-lieutenant or chief lieutenant. This grade was subsequently abolished, and a *new rank* of second captain was instituted. The first was only a lieutenant, *primus inter pares*, while the other was a *captain*, though *en seconde*.

Establishment of remaining 11 (including cadets):—

		£	s.	d.
2 colonels-in-chief and *en seconde*				
1 colonel-commandant		1	5	0
1 lieut.-colonel		1	0	0
1 major		0	15	0
21 captains (captain-lieut. 6s.)		0	10	0
53 subs. 1st lt. 5s., 2nd lt. 4s., fireworker 3s.				
1 chaplain		0	6	8
1 adjutant		0	5	0
1 quarter-master		0	6	0
1 bridge-master		0	5	0
2 medical officers				
48 gentlemen cadets		0	1	4
140 non-commissioned officers	serjeants	0	1	7¾
	corporals	0	1	6½
	bombardiers	0	1	4½
200 gunners		0	1	1¼
616 matrosses		0	0	9½
21 drummers, drum-major	(no drummers to cadet company)	0	1	1¼
fife do.		0	1	1¼
drummer		0	0	9½
1,110 Total				

1749. During the peace the Royal Regiment of Artillery, consisting of ten companies, continued on the same establishment as last mentioned, and were stationed as follows, viz: in Minorca, Gibraltar, Newfoundland, and Scotland, one company at each; the company heretofore at Louisburg, was, on the giving back of that fortress to the French in 1749, sent to Nova Scotia, a new colony, and the remaining five companies stationed at Woolwich, one of which was generally at Greenwich for the ease of quarters.

Distribution:—

1 Gibraltar.
1 Minorca.
1 Newfoundland.
1 America (Nova Scotia).
7 in Great Britain (including cadets).

11 including cadets.

On occasion of the peace of Aix-la-Chapelle, a magnificent firework was exhibited in the Green Park, and the corps of artillery was then reviewed for the first time by the King.

1749. 27 March. Garrison Order by Colonel Belford.—"That the captains provide their men with one pair of white stockings each, in which they are to parade at all times when not under arms."

3 April. Garrison Order by Colonel Belford.—"The non-commissioned officers and men are to be forthwith provided with one good pair of white spatterdashes with black buttons, also with one pair of black spatterdashes, and a rose to their hair, the same as we had in Flanders."[1] The Royal Artillery was the first corps in the British army who wore black spatterdashes.

The two companies in Minorca and Gibraltar were this year relieved from Woolwich for the first time since these places came into the possession of Britian.

6 July. On occasion of the death of the late Master-General the Duke of Montague, the King signed a commission constituting General Sir John Legonier, then Lieut.-General of the Ordnance, to be colonel *en seconde* of the Royal Regiment of Artillery, and captain of the company of gentlemen-cadets.

1750. The regimentals of the non-commissioned officers and men underwent an alteration. The sergeants' coats were laced round the button holes with gold looping; the corporals, bombardiers, and rank and file had yellow worsted looping in the same manner. The corporals and bombardiers had gold and worsted shoulder-knots, the surtouts were laid aside, and complete suits of clothing were delivered yearly. Until 1802 the sergeants were distinguished by gold lace shoulder-knots, a corporal by two worsted ditto, and a bombardier by one ditto, when in place thereof, the sergeants had three gold chevrons, corporals two worsted lace, and bombardiers one on their right arms.

Garrison Order by Colonel Belford.—"The officers are for the future to mount guard in their regimentals, with their fuzees and cartouch boxes."

A captain-lieutenant added to the cadet company, and the first lieutenant reduced.

Establishment as in 1748.

---

[1] Extracted from Kane's List.

1751. Establishment as in 1750.

March. This year Lieut.-General Borgard died. He was the first colonel of the Regiment of Artillery in Britain, and he formed and founded the corps as well as the science of artillery in the British dominions; he was succeeded by Colonel Belford.

30 April. This year the King was pleased to issue a declaration and order under his sign-manual, ascertaining the rank of the officers of the Royal Regiment of Artillery to be the same as that of the other officers of his army of the same rank, notwithstanding their commissions have been hitherto signed by the master-general, the lieutenant-general, or principal officers of the ordnance, which had been the practice hitherto, and from this period all commissions of officers in his Royal Regiment of Artillery have been signed by his Majesty, and countersigned by the master-general of the ordnance.[1]

Hitherto all non-commissioned officers, privates, and even drummers, had warrants signed by the master-general and countersigned by his secretary, appointing such a one a sergeant, bombardier, gunner, or matross, or drummer in the Royal Regiment of Artillery, for which a sergeant paid , a matross or drummer paid , and the other stations in proportion; a custom that with great propriety was now abolished, as no one purpose appears to have been answered by it.

The majority of the officers of the regiment entered into an agreement for the establishing a fund for the benefit of the widows of the corps, no sort of provision having been made hitherto for the widows of the officers in the Royal Regiment of Artillery.

5 April. Garrison Order by Colonel Belford.—"The officers, when on duty, are to wear the knots of their sashes in their left pocket, with the fringe hanging out."

1752. Establishment as in 1750.

3 August. The Duke of Cumberland reviewed the regiment at Woolwich.

---

1 This originated in a dispute at Gibraltar, where the precedence of the Royal Artillery as a corps, and the rank of the field officers of artillery, had been disallowed by the general officer commanding.

Distribution of gunners in Great Britain :—

	Master Gunners.	Gunners.
Berwick	1	6
Blackness Castle	1	1
Calshot Castle	1	2
Carlisle...	1	3
Chester	1	2
Clifford's Fort	1	2
Dover		
Dumbarton	—	1
Dartmouth	1	2
Edinburgh	1	3
Gravesend and Tilbury	2	12
Guernsey	1	4
Hull	1	6
Hurst	1	4
Holy Island	1	1
Inverness	1	2
Jersey	1	8
Landguard Fort	1	3
St. Maurice	1	1
Pendennis	1	2
Plymouth and St. Nicholas	2	18
Portland	1	2
Portsmouth	1	15
Blockhouse Fort	2	1
Charles Fort	1	1
Borough Fort	1	1
Southsea Castle	1	1
Sheerness	1	16
Selby	1	6
Scarborough	1	1
Stirling	1	3
Tower of London	1	4
Upnor Castle	1	12
Cockhamwood	1	1
Gillingham	1	1
Horness	1	0
Fort William	1	2
Windsor	1	1
North Yarmouth	1	2
Sandown, I. of W.	1	3
Fort Augustus...	1	2
Yarmouth, I. of W.	1	5
Carisbrook	1	3
Cowes	1	5
St. James's Park	1	8
	46	179

1753. **Establishment as in 1750.**

On the death of Colonel Thomas Patterson, who had retired from the regiment (as noticed in 1748) on his lieut.-colonel's pay, the lieut.-fire-workers, supported by the Duke of Cumberland and Sir John Ligonier (then lieut.-general of the ordnance), presented a memorial to the king, requesting that his pay might be applied to the making up theirs equal to that of an ensign at 3s. 8d. per day; this was granted, and took place in March this year, their subsistence being now 2s. 9d. per diem and 11d. per diem arrears, instead of 2s. 3d. per diem, and 9d. arrears, which they had hitherto.

1753.
13 June. The king reviewed in the Green Park five companies of the regiment then at Woolwich, and the cadet company. The detail of their strength in the field was:—

1 colonel.
1 lieut.-colonel.
1 major.
5 captains.
6 captain-lieutenants.
4 first lieutenants.
7 second lieutenants.
17 lieutenant-fireworkers.
1 chaplain.
1 adjutant.
1 quarter-master.
1 bridge-master.
1 surgeon.
1 surgeon's mate.

---

15 serjeants.
15 corporals.
1 drum-major.
10 drummers.
6 fifers (one fifer major).
40 bombardiers.
48 cadets.
98 gunners
291 matrosses.

477 Total

The cadets formed on the right of the battalion as grenadiers, and there were three light 6-prs. on each flank of the battalion.

1754. Establishment as in 1750.

March. A detachment, commanded by a captain-lieutenant (W. Hislop), with five officers, twelve cadets, and about sixty non-commissioned officers and privates, embarked for the East Indies.

The halberts were taken from the corporals and bombardiers, and they fell into the ranks with carbines.

1754. 16 July. General Order.—"No non-commissioned officer or private soldier is to appear in ruffles at the review."

During the peace from 1748 to 1756, the Duke of Cumberland reviewed the corps annually at Woolwich.

Nov. This year a detachment of fifty men of the regiment of artillery embarked with the two battalions, No. 44 and 48, under the command of Major-General Braddock, for America. Captain Ord being then with his company in Newfoundland, was by desire of the Duke of Cumberland ordered from thence to command the detachment, which was mostly cut to pieces near Fort Duquesne on the Monongahela, on the 9th July, 1755.

The American war may be said to have commenced from the date of General Braddock's arrival in America in February, 1755. General Braddock marched with the best of his troops—about 2000—from Alexandria (in Virginia) on the 10th June, and on the 8th July he arrived within ten miles of Fort Duquesne on the Ohio.

The whole distance from Alexandria was about 226 miles, but the difficulties of transport were great.

The French were fully prepared to receive Braddock, and attacked him suddenly when on the march on the 8th July. The British lost their commander, twenty officers, and 200 men killed, and twenty-seven officers, and 400 wounded.

This was the first detachment of the corps sent to America; and it was on this occasion that the officers of the Royal Regiment of Artillery serving in the field received for the first time *bât*[1]*-money*, or an allowance *in cash* for the carriage of their baggage; for, as neither horses or carriages could be contracted for, nor procured in any manner in the country which was the scene of Major-General Braddock's operations, Captain Ord, commanding the artillery in that service, ordered each officer to be paid in cash a sum equal to what was allowed by the Board of Ordnance in their contract for horses and carriages in the campaigns during the last war in Flanders, when a captain was allowed a waggon with four horses for the transport of his baggage, and

---

[1] *Batta.*—Here we find the origin of the well known word batta. As used in India it really signifies a money payment made to troops to make good losses which they may have incurred during a campaign, especially losses of bât animals, which officers in India provide for themselves. From the compensation for *bât animals*, it gradually came to signify a money payment in lieu of other allowances which are ordinarily made "in kind." Thus, in India, batta used to signify "*allowances*" as distinguished from *pay* proper, and included payments *in lieu of* house rent, travelling expenses, camp equipment, servants, and rations. The money payment which, until very recent times, was made to the troops in India on Christmas Day, New Year's Day, the queen's birthday, and after the general's inspection, was emphatically called *dry batta* on the *lucus a non lucendo* principle, because it represented *drink*, to which by custom the troops were entitled.

the same to two subalterns, and so in proportion to the other ranks, which has been continued ever since in America (1783).

List of the officers of the Royal Regiment of Artillery, as they stood in February, 1755:—

Colonel ... ... William Belford.
Lieut.-colonel ... ... Borgard Mitchelson.
Major ... ... George Williamson.

Captains:
1. James Mace (Minorca).
2. John Chalmers (Woolwich).
3. Thomas Desaguliers "
4. Thomas Flight "
5. Thomas Ord (Newfoundland).
6. James Pattison (N. Britain).
7. Alexander Leith (Gibraltar).
8. John Skeddy (Woolwich).
9. Charles Brome (Nova Scotia).
10. John Godwin (Woolwich).

Captain-lieutenants:
1. Samuel Shepardson.
2. Edward Sibbet.
3. John Farquaharson.
4. Thomas James (Gibraltar).
5. William Hislop (East Indies).
6. Robert Hind (America).
7. Charles Farrington.
8. Jacob Gregory (Newfoundland).
9. Joseph Brome (Adjutant).
10. John Notshal (Minorca).

First lieutenants:
1. Abraham Tovey.
2. Charles Stranover.
3. Richard Mason.
4. John Dawes.
5. Robert Smith.
6. Leonard Pattison.
7. William Hussey.
8. Samuel Strachy.
9. William Martin.
10. Nath. Jones.

Second lieutenants:
1. John Straton.
2. John Innes.
3. Thomas Smith.
4. W. Macleod.
5. Peter James.
6. Alexander Campbell.
7. Jacob Tovey.
8. Thomas Pike.
9. Joseph Winter.
10. Ed. Whilmore.

Lieut.-fireworkers:
1. Forbes Macbean.
2. Griffith Williams.
3. F. J. Buchanan.
4. John Macculloch.
5. Sir Chas. Chalmers.
6. George Charleton.
7. Robert Hind.
8. Joseph Banet.
9. Richard Edlin.
10. Thomas Harvey.
11. D. Hay.
12. James Stephens.
13. W. Phillips.
14. J. Yorke.
15. G. Anderson.
16. A. Ferguson.
17. D. Rogers.
18. H. Spry.
19. B. Stenelen.
20. D. Drummond.
21. J. Lawson.
22. C. Torriano.
23. T. Howdell.
24. J. Butler.
25. D. Ninckle.
26. G. Foreman.
27. James Hulsil.
28. J. Mullman.
29. Thomas Baker.
30. Sir John Withwronge

Staff officers:
Chaplain—Montagu Barton.
Adjutant—Joseph Browne.
Quarter-master—W. Phillips.
Bridge-master—Nath. Marsh.
Surgeon—James Irwine.
Mate—David Vaus.

1755. March. Four companies, with an additional major, were added to the regiment, which now consisted of fourteen companies, the cadets included.

These four companies were raised and completed in about thirty days, and immediately embarked for the East Indies; one of them was lost in the "Doddington" East Indiaman on her passage, all except three private men; as soon as this was known at home a company was raised to replace it, and embarked for the East Indies.

The "Doddington" sailed from the Downs, April 23rd, 1755, in company with the "Pelham," "Houghton," "Streatham," and "Edgecourt," all in the service of the East India Company. On the 17th July, 1755, the "Doddington" struck the shore about 250 leagues east of the Cape of Good Hope, in latitude 33° 44′ South, on a barren, uninhabited rock. Of 270 souls that were on board when she struck, only twenty-three reached the shore, including three matrosses, viz., John Lister, Ralph Smith, and Edward Dyroy. On this rock they remained exactly seven months, having suffered the pangs of hunger during that period. They finally contrived to rig a boat, which they called the "Happy Deliverance," and launched on 16th February, 1756. On the following day they got their scanty supply of stores on board, and on the 18th February set sail from the rock on which they had lived just seven months, and to which they gave the name of "Bird Island," for the island abounded with birds.

For a detailed account of this wreck, vide "Ann. Reg.," p. 287-297.

1755. May. A detachment of a first lieutenant and twenty-four non-commissioned officers and private men were this year sent from Woolwich to Ireland, at the request of the lord lieutenant, towards the forming a battalion of artillery for that kingdom. Captain John Stratton appears to have been the first captain of the Irish Artillery, and was commissioned as captain in April, 1756.

Oct. Two more companies were raised and added to the regiment, which now consisted of sixteen companies, exclusive of the cadets.

1756. 13 Feb. Geriah Fort, in the East Indies, taken by Admiral Watson. Captain Tovey, with a detachment of Royal Artillery, did good service with his four bomb vessels.

21 Feb. Capt.-Lieut. Strachey ordered to join the artillery in the East Indies.

1 April. Artillery Company of Ireland:—

Major H. Brownrig.
Capt. Jas. Stratton.
First-Lieutenant G. Skipton.
Second-Lieutenant W. Gray.
Lieutenant-Fireworkers { W. Brady. J. Ratcliff. R. Graham.

5 sergeants.
5 corporals.
106 bombardiers.
34 gunners.
102 matrosses.
2 drummers.

From the Army List of 1758.

1756. 1 April. Two companies were raised and added to the regiment, which now consisted of eighteen companies, exclusive of cadets.

Forbes Macbean appointed adjutant to the regiment.

15 April. "It is the Duke of Cumberland's orders that Colonel Bedford write to Captain Patterson to acquaint General Bland it is his royal highness's commands that the artillery take the right of all foot on all parades, and likewise of dragoons when dismounted."[1]

16 April. Minorca besieged by the French. Captain Flight commanded the artillery, consisting of one company. Lieutenants Charlton and Stehelin present with it; other officers not known. Lieutenant Charlton was adjutant at the siege.

12 May. A company of miners, consisting of 200, was raised for the purpose of being sent to assist in retaking Minorca. First Lieutenant William Phillips, then the youngest of that rank in the regiment,[2] being aide-de-camp to Sir John Ligonier, was, through that interest and the indolence and supineness of the officers of the corps, appointed captain of the company, which was then considered as no more than a temporary affair of short duration, that would end when the service they were raised for was performed; but on the return of this company (1757) to Woolwich, the miners who did not choose to remain as privates in the regiment were discharged, and the company added to, and established in, the regiment on the same footing as the others. This made nineteen companies besides the cadets.

---

1 In a report on the "Precedence of the Royal Artillery," signed by the colonels commandant of the several battalions, in the year 1783, on the occasion of an appeal to the king from the officer commanding the royal artillery at Gibraltar, the origin of this order of the Duke of Cumberland is thus given:—"Nor can we, my lord, recollect but a single instance where this privilege has been called in question. This happened at Perth, in the year 1756, when all the troops quartered in that town were ordered to be present upon the South Inch at the execution of a deserter, upon which occasion a dispute arose between Major East, commanding the Queen's Dragoon Guards, and Major-General Pattison, then commanding a company of artillery stationed there, which of the two corps should take the right on the field, and, neither officer being willing to yield up the point, the matter was composed for that day without the right being ceded by either party, reserving it to be decided by authority. The matter was therefore immediately referred to H.R.H. the Duke of Cumberland, at that time commander-in-chief of the army, and his royal highness was therefore pleased to issue his order confirming the privilege in favour of the artillery, as will appear by the enclosed extract from the regimental order of the 15th April, 1756, and a memorandum similar to it found amongst the papers of the late adjutant-general, and given by Lieut.-General Fawcet to Major-General Brome."

2 The commissions of William Phillips were dated:—First lieutenant, April 1, '56; captain, May 12, '56: brevet lieut.-colonel, August 15, '60; major, April 25, '77; lieut.-colonel, July 6, '80. He died in Virginia, May 13, 1781.

1756. 1 June. John Pomeroy appointed quartermaster to the regiment.

June. A great train of artillery, with five companies of the regiment and the company of miners, marched from the Tower of London, and encamped at Byfleet, in Surrey, till October.

The guns comprising this train were the following:—

Twenty light 6-pounders.
Eleven „ 24-pounders.
Six „ 3-pounders.
Fourteen „ 12-pounders.
Six royal mortars, and
Ten pontoons.
TOTAL 57.

The Duke of Marlborough, then master-general, marched at the head of this train, and pitched his tent in camp, where his grace remained during the campaign, attending various experiments in mining, and the usual exercises with the guns and mortars, &c.

The following details refer to the encampment at Byfleet, in Surrey:—

June. A train of artillery encamped at Byfleet, in Surrey.

Ordnance light brass, mounted on travelling carriages, ammunition boxes and elevating screws	24-pounders	11
	12 „	14
	6 „	20
	3 „	6
	Royal Mortars	6
Round shot fixed to wood bottoms, flannel cartridges filled, and parchment covers.	24-pounders	550
	12 „	700
	6 „	2,000
	3 „	600
Shells empty	$5\frac{1}{2}$ howitzers	300

List of the staff officers, officers, ministers, and attendants appointed to attend the field train of artillery, with their respective pay:—

		Per diem. £	s.	d.
Master-general	Duke of Marlborough	10	0	0
Lieut.-general	Rt. Hon. Sir John Ligonier	3	0	0
Aide-de-camp to master-general	Thos. Desaguliers	0	10	0
Secretary to do.	Jacob Bryant	0	10	0
Aide-de-camp to lieut.-general	William Phillips	0	10	0
Clerk to do.	John Hayton	0	10	0

		Per diem. £	s.	d.
Comptroller	Thomas Jones	2	0	0
Chaplain	Anthony Sylk	0	6	8
Commissary of horse	George Williamson	0	10	0
Commissary of stores and paymaster	John Barnes	0	8	0
1st clerk of stores	Wm. Saltonstall	0	4	0
2nd do.	Thos. Bowden	0	3	6
1st conductor	Thos. Hodgson	0	3	0
2nd do.	Thos. Adams	0	3	0
3rd do.	Hugh Griffiths	0	3	0
4th do.	Wm. Forman	0	3	0
5th do.	Thos. Bryce	0	3	0
6th do.	John Gentle	0	3	0
7th do.	William Alcock	0	3	0
8th do.	Richd. Gardner	0	3	0
9th do.	Richd. Heydon	0	3	0
10th do. messenger to the master-general	James Westall	0	3	0
11th do.	Arthur Neil	0	3	0
Kettle-drummer	Cotterel Barret	0	3	0
Kettle-drum driver	Wm. Wakefield	0	3	0
1st carpenter	John Hackney	0	3	0
2nd do.	George Crips	0	3	0
1st smith	James Coggle	0	3	0
2nd do.	Alexr. Coggle	0	3	0
1st wheelwright	Daniel Barry	0	3	0
2nd do.	John Parker	0	3	0
1st collar maker	John Whitaker	0	3	0
2nd do.	John Workworth	0	3	0
3rd do.	Willm. Loose	0	3	0
Tinman	Francis Perrins	0	3	0
Cooper	John Greenwood	0	3	0

Artillery.

			Regimental pay. £	s.	d.
Colonel commandant	William Belford		1	5	0
1 major	Geo. Williamson		0	15	0
7 captains		each	0	10	0
5 captain lieutenants		do.	0	6	0
6 first lieutenants		do.	0	5	0
2 second lieutenants		do.	0	4	0
1 lieutenant miner			0	3	8
16 lieutenant fireworkers		each	0	3	8
18 sergeants		do.	0	2	0
19 corporals		do.	0	1	10
39 bombardiers		do.	0	1	8
70 gunners		do.	0	1	4
350 mattrosses		do.	0	1	0
80 miners		do.	0	1	9
13 drummers		do.	0	1	0

1756. Distribution of waggons, drivers, and horses, &c., as they marched out of the Tower with the train, for the camp at Byfleet, the 12th July :—

		Waggons. Carts, &c.	Drivers. To each.	Drivers. Total.	Horses. To each.	Horses. Total.
Kettle-drum carriages	1	—	—	1	Greys.	6
Brass ordnance. 24-pounders	1	with	flag.	2	—	6
	10	—	2	20	5	50
12 "	14	—	1	14	3	42
6 "	20	—	1	20	2	40
3 "	6	—	1	6	2	12
Howitzers, brass, 5½ inch ...	6	—	1	6	2	12
Spare carriages. 12-pounder ...	1	—	—	1	—	3
6 " ...	1	—	—	1	—	2
Pontoons	1	with	flag.	2	—	6
	9	—	2	18	5	45
Spare carriage for pontoon ...	1	—	—	2	—	5
For ammunition and stores. Wagns.		1 with	flag.	2	—	5
		42	1	42	4	168
Ammunition carts		22	1	22	3	66
Tumbrils		2	1	2	2	4
Forge carts		1	—	1	—	2
For triangle gyn, park pickets, laboratory tent ... Wagns		2	1	2	4	8
For the intrenching tools ... "		2	1	2	4	8
Master-general ... ... "		6	2	12	5	30
Did not march. Chaise marine...		1	—	1	—	4
Lieut.-general ... ... Wagns.		5	2	10	5	25
Chaise marine		1	—	1	—	4
Surveyor-general ... Wagns		3	2	6	5	15
Did not march. Chaise marine...		1	—	1	—	4
For the colonel commandant Wagns.		2	1	2	4	8
Major and commissary of horse "		1	—	1	—	4
Master-general's secretary ... "		1	—	1	—	4
Comptroller ... ... "		1	—	1	—	4
5 captains, 6 captain-lieutenants, 1 chaplain ... ... "		4	1	4	4	16
28 lieutenants ... . "		7	1	7	4	28
Surgeon and mate ... ... "		1	—	1	—	4
Quartermaster, adjutant, bridge-master ... ... "		1	—	1	—	4
Commissary and pay-master, 2 clerks of stores ... ... "		1	—	1	—	4
6 companies of artillery ... "		6	1	6	4	24
10 conductors ... ... "		1	—	1	—	4
10 artificers and their tools ... "		2	1	2	4	8
TOTAL ...				225		684

The order in which the train marched out of the Tower.

2 tumbrils with intrenching tools.
Kettle-drums.
1 light 24-pounder with a flag.
10 24-pounders.
14 12-pounders.
20 6-pounders.
6 3-pounders.
6 howitzers.
1 12-pounder } Spare carriages.
1 6 ,, }
22 ammunition carts.
The flag waggon.
6 waggons for the companies.
40 waggons.
The gyn and laboratory tent waggons.
2 waggons with intrenching tools.
3 spare waggons.
5 waggons for the lieut.-general.
2 waggons for the colonel commandant.
1 waggon for the major.
4 waggons for the 5 captains, 6 captain-lieutenants, and chaplain.
7 waggons for 28 lieutenants.
1 waggon for the surgeon and mate.
1 waggon for the quarter-master, adjutant, and bridge-master.
1 waggon for the comptroller.
1 waggon for the commissary, and 2 clerks or stores.
1 waggon for 10 conductors.
2 waggons for 10 artificers and their tools.
2 waggons for the driver's tents.
Forge cart.
1 pontoon with flag.
9 pontoons.
1 spare pontoon carriage.

FORAGE allowed to the staff officers and others attending the field train which marched out of the Tower the 12th July, decamped the 20th of October, 1756.

200 days to each at their first encamping, and 100 days more before they decamped; 300 days in all.

	Rations per day.
Master-general	40
Lieut.-general	30
Colonel commandant of artillery	10
Major of artillery	6
Aide-de-camp to the master-general	4
Comptroller to the train	8
Captains, and captains-lieutenants of artillery each	4

First and second lieutenants, and lieutenant fireworkers ... ... each	1	At 6d. per day each.
Commissary of horse	4	
Commissary of stores ... 4 / And as paymaster ... 4	8	
Chaplain	2	
Adjutant	1	
Quartermaster	1	
Bridge-master	1	
Practitioner engineers ... ... each	1	
Surgeon	2	
Surgeon's mate	1	
Assistant paymaster	2	
Clerk of stores	1	

N.B. The commissary of horse, master-general's aide-de-camp, adjutant, and quartermaster, receive their allowances likewise as officers of artillery, according to their rank.

Wood and Straw.—To each tent, 5 trusses of straw the first week, and 1 truss per week for each tent afterwards; and 24 lbs. of wood per day for each mess, or every two tents, for dressing their meat.

Instructions for Thomas Jones, Esq., comptroller of the train in Great Britain:—

1st. You are to muster the officers, ministers, and attendants belonging to the train, both civil and military, at such times and places as shall be ordered by the master-general, or in his absence by the lieutenant-general, always taking care that no persons are mustered but such as are or may be ordered by the master-general, and to cause pay lists to be made out from time to time accordingly; and you are regularly to transmit such muster rolls to us, signed by you, and attested by the commissary of stores.

2nd. You are likewise, in the absence of the surveyor-general, to muster the drivers and horses once every month, or oftener if necessary, and transmit to us a certificate, signed by yourself and the commissary of horse, of the number of the drivers and horses so mustered, remarking in the said certificate if any of the horses are lamed or otherwise unfit for the service; and you are to send a duplicate of the said certificate to the surveyor-general in order that he may muster and mark other horses in lieu of such as are lame or otherwise unfit for the service.

3rd. You are to report to the surveyor-general once a month or oftener the condition of the waggons, ammunition carts, harness, &c., to the intent that he may give you the necessary orders for repairing and replenishing whatever may be judged necessary.

4th. You are once every month, or oftener if you see it necessary, to demand a state of cash from the paymaster, with an account of all receipts and payments made in the preceding month, to the end you may have a proper checque upon him. You are to keep a book in which shall be entered all orders for payments of money, whether given by the master-general, or, in his absence, the lieut.-general, before the same be sent to the paymaster for payment.

5th. If it should hereafter be found necessary that the paymaster should draw on this office, you are in that case to countersign his draught, and you will observe that all bills of incidental charges are to be comptrolled and signed by you, and then you are to present them to the master-general, or, in his absence, the lieut.-general, who will order them for payment if they judge the same to be reasonable; but whenever the service shall be so pressing as to require the immediate payment of any incidental charges, you may in such cases only give the order for payment yourself, and afterwards obtain the master-general's approbation thereof in writing, which must accompany the paymaster's accounts when they are transmitted to the surveyor-general's office.

6th. You are to make timely demands of whatever stores, materials, straw, firing, and whatever may be wanting for this train, and at the same time to signify to us whether the whole or any part of your said demand can be furnished upon the spot, and at what rates, that we may consider of the best and cheapest of complying with your demands; and you will take notice that you are not to suffer any stores, materials, forage, straw, &c., to be bought or paid for without orders in writing from the master-general, lieut.-general, or us.

7th. You are to inspect the commissary of stores' accounts, to see that his journal of stores received and issued be fairly and constantly kept up according to his instructions.

8th. You are to obtain orders in writing from the general or the officer commanding in chief for all ammunition or stores, to be issued to the troops or otherwise, and from the master-general or lieut.-general for all such as are for the service or use of the train; and to see that the commissary of stores does not make any issues but by such orders.

9th. You are to see that the commissaries, clerks, conductors, and artificers do perform their respective duties, as also that the commissary of horse reports to you, as often as you see necessary, what horses are disabled, killed, or die in the service; and that he give you a state of the harness that may want repairing or are not in a condition for the service, that you may thereby be enabled to make the surveyor-general the return before required.

10th. You are to direct each artificer to keep an account of what work he does from day to day, and of the materials expended on each particular service, which accounts are to be delivered to the commissary of stores or his assistant, who are to make two copies thereof, one to be sent to the surveyor-general, the other to be delivered to you.

11th. If any commissary, clerk, conductor, or driver neglect their duty, or are guilty of any crime or misbehaviour, they may be confined to their tents or quarters, but not sent to the common gaol or guard, unless any particular circumstance of the misdemeanour requires it; and you are to report the same to the master-general or lieut.-general, with the nature of the offence, for their further orders; and no person so confined shall be mulcted of his pay or any part thereof, or be dismissed the service, but by sentence of a general court-martial or by the order of the master-general.

12th. If any quarrels or disputes arise by the ill conduct of the contractors or their drivers with the civil or military officers, the same shall be reported by you to the master-general, or, in his absence, the lieut.-general, for their determination thereon.

Lastly. You are carefully and diligently to discharge the duty of comptroller to the artillery, by doing all things thereto belonging, and follow such orders and directions as you shall from time to time receive from the master-general, lieut.-general, and principal officers of his majesty's ordnance, and the general, or commander-in-chief of his majesty's forces for the time being.

Given at the office of his majesty's ordnance, under our hands and the seal of the said office, the thirty-first day of July, 1756.

J. L. Ligonier.
Charles Frederick.
W. R. Earle.

Instructions for Major George Williamson, commissary of horse to the train of artillery in Great Britain:—

1st. You are to take care that all the horses which are or may be provided for service of the said train, be well looked after, kept constantly clean, shod, and in good order and condition, likewise that the drivers do diligently and faithfully perform their respective duties.

2nd. You are to be present when the comptroller musters the horses, in the absence of the surveyor-general, and to join with the comptroller in giving a certificate of the number of drivers and horses mustered, in which certificate it must be remarked whether any of the horses are lame, or otherwise unfit for the service, and a duplicate of the said certificate is to be transmitted to the surveyor-general in order that he may muster and mark other horses in lieu of such as are lame or otherwise unfit for the service.

3rd. You are to take care that none of the king's harness, tools, materials, or stores belonging to the horses or waggons are misapplied, embezzled, or converted to private use, on any pretence whatever.

4th. Whereas we have contracted with Messrs. Warrington and Baldwin that none of the horses provided by them shall be employed in carrying provisions, or any other materials which shall not be for the immediate use of the train, without licence from the commander-in-chief:—You are therefore to be particularly careful not to suffer any of the horses or drivers to be employed for any other use or service than those for which they are hired, without orders in writing from the master-general.

5th. You are once every month, or oftener if required, to lay before the comptroller an account of what horses are killed, or die in the service, or are disabled; as also a state of the harness, &c., that may want repairing, or are not in a condition fit for service.

6th. When the train marches, you are to see that all the carriages, tumbrils, and waggons proceed according to the disposition that shall be given you by the master-general, or, in his absence, the lieut.-general, and that the several attendants thereon be in their proper posts and divisions.

LASTLY. You are to give a copy of these instructions to the comptroller, and strictly to observe the same yourself, as you will answer the contrary.

Given at the office of ordnance under our hands and the seal of the said office this thirty-first day of July, 1756.

J. L. LIGONIER.
CHARLES FREDERICK.
WM. R. EARLE.

1756. 28 June. Minorca capitulated; the garrison sent to Gibraltar.

The Board of Ordnance ordered a half a day's pay in addition to each of the N. C. officers, gunners, mattrosses, and drummers, during the seige of St. Philip's Castle, at Minorca, viz., from 18th April to the 28th June, 1756, both days included. Also to the following officers, under Captain Flight's command:—

Captain	Thomas Flight.
Captain-lieutenants ...	Jacob Gregory. Philip Webdell. George Charlton.
First lieutenants ...	George Charlton, promoted to captain-lieutenant. Benjamin Stehelin.
Second lieutenants ...	Benjamin Stehelin, promoted to first lieutenant. George Foreman. David Day.

The above had half a day's pay each from 29th June to the 23rd July, on account of extra duty at the siege of Minorca, granted by the board.

1 Aug. The troops from Minorca landed at Gibraltar.

2 Aug. Captain Flight's company with the other troops commenced garrison duty.

Strength of the artillery when besieged:—

2 captains.	11 corporals or bombardiers.
5 subalterns.	85 privates.
1 surgeon.	2 drummers.
3 sergeants.	

Killed............... 14 }
Wounded .... 19 } during the siege.

1756. 4 Aug. Byfleet camp. Orders by Colonel Belford. Disposition of the officers commanding the several brigades.

24-pounders	1st brigade	4 guns	Captain Godwin.
	2nd "	4 "	Major Cleaveland.
	3rd "	3 "	Captain Brome.
12-pounders	1st brigade	5 guns	Captain Hussey.
	2nd "	5 "	Captain Strachey.
	3rd "	4 "	Captain Smith.
6-pounders	1st brigade	5 guns	Captain Hay.
	2nd "	5 "	Captain Stephens.
	3rd "	5 "	Lieutenant York.
	4th "	5 "	Lieutenant Anderson.
3-pounders		6 guns	Lieutenant Drummond.
Royal howitzers	1st	3 hwtzrs.	Captain Tovey.
	2nd	3 "	Lieutenant Tornant.

31 Aug. America.—Frontinac taken by Colonel Bradstreet with a detachment of troops and 27 of the Royal Artillery.

15 Sept. Regimental order at Byfleet Camp, by Colonel Belford:—"The non-commissioned officers and men are on all occasions for the future to wear their hair clubbed."

26 Oct. Lieutenant-Fireworkers William Hill, William Harris, Robert Rogers, and William Lee ordered forthwith on board a transport for Gibraltar.

Nov. A company embarked for Gibraltar, in addition to the one already there.

A company embarked for the East Indies to replace that lost in the "Doddington."

1757. 4 Feb. The regiment consisted of nineteen companies.

Borgard Mitchelson	Appointed	colonel commandant.
George Williamson	"	lieutenant-colonel.
Thomas Desaguliers	"	" "
John Chalmer	"	major.
Thomas Flight	"	"
Jacob Gregory	"	captain.
Samuel Strachey	"	"
Dun. Drummond	"	adjutant.
Thomas Baker	"	quarter-master.
George Anderson	"	bridge-master.

11 Feb. Lieutenant-Colonel Williamson received the master-general's orders and instructions for commanding and embarking the artillery on an expedition to North America, under the Earl of Loudoun. Captains Godwin's and Strachey's companies embarked at Woolwich

for this purpose, Captain-Lieutenant Stephens, Lieutenants Torriano and Walton getting ready for the same service in bomb vessels.

The following officers embarked with the companies :—

Captains	Godwin. Strachey
Captain-lieutenants	Smith. Hay.
Second lieutenants	George Lewis. Thomas Blomer. Joseph Walton. Thomas Shepherd. Fleming Martin. Samuel Glegg.
Lieutenant-fireworkers	Henry Skinner. J. Williamson. John Wilson. Peter Traill. David Standish. William Johnstone. Thomas Simpson. John Stewart.

1757. Feb. Two companies of the regiment, under the command of Lieut.-Colonel Williamson, embarked for America, to serve with the army forming there by Lieut.-General the Earl of Loudoun.

21 Feb. Lieutenant-Colonel Williamson, with the artillery, arrived in the Downs.

4 March. Lieutenant-Colonel Williamson, with the artillery, arrived at Spithead.

16 April. The expedition sailed from Spithead.

25 April. The expedition arrived at Cork.

2 May. Captain-Lieutenant Webdal arrived at Cork, and relieved Captain Smith, promoted, who disembarked.

8 May. The expedition sailed from Cork.

8 July. The expedition arrive at Halifax. Captain-Lieutenant Buchannon, Lieutenants Howdell and Maine with a detachment, join Colonel Williamson.

9 July. The army divided into three brigades, with Lieutenant-Colonel Williamson and 370 artillery men, and prepare for the attack of Louisburg.

This force joined Lord Loudoun's army at Halifax, and consisted of twelve regular battalions (forming three infantry brigades) and a detachment of 400 artillerymen. The fleet to co-operate consisted of sixteen line of battle ships and eight frigates. The expedition was bound for Louisburg (Cape Breton), and when on the point of starting, on the 4th August, intelligence was received of a very superior French force having arrived at Louisburg from France, and that the garrison consisted of 3,000 regular troops. The expedition was given up, and the troops were divided and placed in garrison at Halifax, at Annipolis, and Cumberland Fort (Beausejour); the remainder went with Lord Loudoun to New York, and thence to Albany.

1757. By Colonel Belford, Woolwich. It is Colonel Belford's
13 Feb. orders that the cadets do on all occasions pay the due respect to the commissioned officers. The first one that is known to pass an officer without pulling off his hat, shall be sent into confinement.

14 April. Woolwich.—Four companies added to the regiment, making twenty-four, including the cadets and miners.

Promotions.—Captain-Lieutenants Dover, Martin, Innis, and Smith, to be captains; Lieutenants Ferguson, Spry, Hind, Stehelin, Drummond, Torrianno, Butler, and Howdell, captain-lieutenants. Dated 2nd April, 1757.

Captain-Lieutenant Spry having accepted a company in the marines, Lieutenant Mollman is promoted to captain-lieutenant.

Men picked out for Captain Hislop's company, for the East Indies, in room of those lost in the "Doddington," to embark at a moment's notice.

28 April. By Colonel Belford. The first cadet that is found swimming in the Thames shall be taken out naked and put in the guard-room.

May. The Corps of Engineers obtained military rank and title for the first time.

The Board of Ordnance ordered the arrears of pay to the following officers of artillery serving in the East Indies:—

Captains	John Skiddy. John Farquharson. Richard Maitland. John Northall. William Hislop.
Captain-lieutenants	Richard Mason. Jacob Tovey. Leonard Pattison.
First lieutenants	Joseph Winter. Edward Whitmore. Sir Charles Chalmers. Joseph Barrett. Thomas Hussey.
Second lieutenants	Jonathan Lewis. David Maskell.

Lieutenant-fireworkers.
- Robert Hatzler.
- George Grove.
- John Chalmers.
- John Smith.
- William Corbet.
- Daniel Sweet.
- John Scott.
- James Wood.
- Alexander Johnson.
- George Campbell.
- Edward James.
- Thomas Saunders.
- David Vans.
- Joseph Brereton.
- George Elliott.

ESTABLISHMENTS of the two battalions of Royal Artillery :—

2 colonels-in-chief *en seconde*.
2 colonels-commandant.
2 lieut.-colonels.
2 majors.
48 captains.
117 subalterns.
1 chaplain.
2 adjutants.
2 quarter-masters.
1 bridge-master.
3 medical officers.
48 gentlemen cadets.
322 non-commissioned officers.
400 gunners.
1,722 matrosses.
47 drummers.

2,531 Total.

1757. 26 June. Disposition for the march of Captain Patterson's company, and the seven detachments to Barham Downs :—

	Captains.	Lieutenants.	N. C. officers.	Gunners.	Mattrosses.	Drums & fifes.
Advanced guard	—	—	1	—	6	—
Front guard	1	2	8	20	40	3
Flanks of 16 guns	—	8	8	16	16	—
Flanks of 1 howitzer ...	—	—	2	—	—	—
Flanks of 10 waggons ...	—	1	4	—	20	—
Rear guard	—	—	4	5	12	—
To act as quartermaster ...	—	1	1	—	—	—
TOTAL	1	12	28	41	94	3

The advanced guard is to march with fixed bayonets and clear the road, and put out of the way fire, or smokers of tobacco, &c.

The gunners to march on the left of their respective guns with their arms slung, and a linstock over their left arm, and the mattrosses to march opposite the gunners with shouldered arms.

1757. 2 July. By Colonel Michaelson. The cadets to be in the academy at 9 in the morning, and at 3 in the afternoon, at which time the officer attending is to see the corporal call the roll, and to report the time they come in after the academy hours, as well as those who are absent.

The officer on duty in the academy is to suffer no disturbance or playing there, and if any cadet shall presume to do so, the officer is to send one of the corporals to the guard for a file of men to take the offender to his rooms; and if any one who attends the academy, and does not belong to the company, shall behave so, the officer is to turn him out.

29 July. A secret expedition (against Rochfort) ordered under Admiral Hawke, and Sir John Mordant, commander-in-chief of the land forces, with the following ordnance and artillery, engineers, with ammunition, stores, &c.

On travelling carriages.	24-pounders	6
	12-pounders	4
	6 light pounders	10
	3-pounders	6
On their beds	10-inch mortars	2
	8-inch mortars	2
	4½-inch mortars	20
On travelling carriages.	5½-inch howitzers	2
	Petards	3

List of the several persons who are to attend the said expedition :—

Artillery.

Captain	Thomas James.
Captain-lieutenant	Peter Innes.
First lieutenant	David Day.
Second lieutenant	Joseph Eyre.
Lieutenant-fireworkers	James Donnellan. James Sawerby. Richard Harrington.

3 sergeants.
3 corporals.
8 bombardiers.
20 gunners.
64 mattrosses.
2 Drummers.

Detachment of Miners.

Lieutenant-fireworker .... Edward Nethercotte.
1 sergeant.
2 corporals
18 miners.

On the Staff, Military Officers.

		Per diem.		
Captain	Thomas James	0	10	0
Adjutant	Lieut. James Donnellan...	0	2	0
Quartermaster	Lieut. Ed. Nethercotte ...	0	2	0
Chief engineer	Robert Clerk	1	0	0
Sub. Engineers	Richard Dudgeon	0	8	0
	Thomas Walker	0	8	0
Practitioner engineers	Robert George Bruce ...	0	7	0
	Augustus Durnford	0	7	0
	William Roy	0	7	0
	Jno. Chris. Eisen	0	7	0

Civil Officers.

Comptroller	Robert Boyd	1	0	0
Clerk of stores and paymaster	Pitman Stagg...............	0	5	0
Surgeon....................................	John Lowndes	0	6	0
6 conductors, each at		0	3	0
2 carpenters, each...........................		0	3	0
2 wheelers, each		0	3	0
2 smiths, each		0	3	0
1 collar maker		0	3	0
1 cooper..		0	3	0

INSTRUCTIONS for Robert Boyd, Esq., comptroller of the artillery.

You are to muster the officers, engineers, and all others belonging to the artillery, both civil and military, once every month, or oftener if necessary, at such time and place as shall be agreed upon between the commanding officer of artillery and yourself, having regard in the said muster to the establishment (a copy whereof is herewith sent you), and noting thereon all such persons as are dead, sick, wounded, deserted, or absent, with or without leave, since the first establishment, or last muster, and accordingly to make out pay lists for their subsistence (according to the enclosed printed forms), to be signed by you and the commanding officer of artillery or engineers, as the same may relate to either of the respective corps, which you will observe are separate and distinct from each other, and you are to take care that no person's name is inserted in any of the said pay lists who is dead, deserted, or absent without leave, except that it appears by the surgeon's certificate that any of the parties are sick who appear to be absent without leave; and you are by the first opportunity to transmit to us copies of the said muster rolls and pay lists, attested by you and the commanding officer of artillery.

You are once every month, or oftener if you see necessary, to demand a state of cash from the paymaster, with an account of all receipts and payments made in the preceding month; and, to the end you may have a proper cheque upon him, you are to keep a book in which shall be entered all orders for payment of any money whatever given by you and the commanding officer of artillery, or commanding engineer, jointly, or by either of you, before the same be presented to the paymaster for payment thereof; and you are to see that the paymaster be supplied with money at the most advantageous exchange, which you are to enquire into carefully, and inform us from time to time.

You are to prevent as much as possible any increase of allowances or payments to any persons whatsoever over and above what the establishment admits of. All incident charges shall be ordered by you and the commanding officer of artillery jointly or separately, as it shall happen, and the paymaster's draughts on the officer to be countersigned in like manner.

You are to inspect the clerk of the stores' accounts, to see that his journals of stores received and issued be fairly and constantly kept up, showing to or from whom and for what service such receipts and issues are made, and that the same be regularly posted from the journals into a ledger, to the end you may at all times readily know the balance or remain of every species of stores, which accounts the commanding officer of artillery may see whenever he shall require the same.

You are to obtain orders in writing from the general or commander-in-chief of his majesty's forces for all ammunition or stores to be issued to the troops or otherwise, except such as are for the service or use of the train; and you are to see that the clerk of the stores doth not make any issues but by orders signed by the commanding officer of the train, which, with the receipts of the persons to whom the issues are made, will be the clerk of the stores' voucher.

You are to act jointly with the commanding officer of artillery in all things relative to stores or cash, and to aid and assist him all you can; but you are by no means to interfere with his command over the officers and men belonging to the Royal Regiment of Artillery employed upon the present service.

If any clerk, conductor, artificer, or other attendant employed upon this service (the regimental corps and engineers excepted) neglect their duty, or are guilty of any crime or misbehaviour, they may be confined to their tents or quarters by the commanding officer or by you, the comptroller, but not sent to the common guard, or personally ill-treated by either; and no person so confined shall be mulcted of his pay or any part thereof, or be dismissed the service, but by the sentence of a general court martial, or by the joint consent of the commanding officer of artillery and you, the comptroller, in which latter case you are to acquaint us therewith immediately, and to give your reasons for so doing.

LASTLY. You are carefully and diligently to discharge the duty of comptroller of artillery, by doing all manner of things thereto belonging, and to follow such orders and directions as you shall from time to time receive from the master-general, lieut.-general, and principal officers of his majesty's ordnance, and the general or commander-in-chief of his majesty's forces for the time being.

(Signed) J. L. LIGONIER.
CHARLES FREDERICK.
W. R. EARLE.

INSTRUCTIONS for Captain Thomas James, commanding officer of the artillery going to North America.

Whereas, you are appointed to command a detachment of the Royal Regiment of Artillery ordered to be embarked on board the transports now in the river Thames:—We think it necessary that you should know with certainty what provision is made for and on account thereof, as also what directions are given to the several persons under your command, and therefore we have directed the clerk of the stores to deliver you as soon as possible a fair copy of the proportion of artillery and stores for the service, distinguishing how the same are distributed on board the several transports. We have also directed the clerk of the stores and paymaster and the surgeon to show you their instructions, and give you copies thereof. You are to see the detachment and the whole establishment (engineers excepted, they being a distinct and separate corps from the artillery) mustered monthly, if time and place will permit, taking care that mention is made in the muster rolls of such as are sick, wounded, deserted, or dead, and according thereto you are to make out and sign the lists of payment for their monthly subsistence, agreeable to the forms hereunto annexed, in which pay lists no person is to be included who is dead, deserted, or absent without leave, unless it shall appear by certificate from the surgeon that such persons are sick who are absent without leave.

If any of the officers or others (engineers excepted) should be killed or die on this service, you are to supply such vacancies for the present by posting the next in rank to act in such vacated stations.

You are to secure the effects of the officers and others (engineers excepted) who are killed or die on this service, and to cause an inventory to be taken of the same as soon as possible, in presence of three or four persons belonging to the artillery, and transmit a duplicate of such inventory to this board by the very first opportunity, so that the lawful representatives of the persons so dying or being killed may be satisfied of your impartiality therein.

You are to require from the paymaster every month, or oftener if you think necessary, a state of his cash, and an account of all receipts and payments for service of the artillery; and because no payments, except what relate to the engineers, are to be made without your orders, you are to keep a fair book thereof as a cheque upon the said paymaster; and that you may know what money he has already received, we inclose an account of what has already been imprested him.

You are not to suffer or permit any increase to be made to the establishment, nor to order the buying or paying for fascines, gabions, or such other things as are always made by the forces and paid for out of the contingencies of the army; nor are you to allow any incident charges but such only as are absolutely necessary for service of the artillery, for which there must be given sufficient reasons to justify the same.

You are from time to time to send us duplicates of the muster rolls, noting therein whatever alterations may happen to be made among the officers and others employed upon this service.

In case the "Fire Drake" or any other bomb vessel should join you on this service, we hereby direct you to take under your care the artillery and stores which they and their tenders may have on board, together with the officers and men thereto belonging; and that you may know what has been done with respect to the "Fire Drake," you will receive herewith a copy of the proportion of stores allotted for them, also copies of the contract for the tender, and of the instructions to Captain-Lieutenant John York, appointed to command the artillery on board the bomb vessel, together with a copy of the instructions to the master of the tender.

LASTLY. You are to follow such orders and directions as you shall receive from time to time from the master-general, lieutenant-general, and principal officers of the ordnance for the time being, the commander-in chief of the forces ordered upon the present expedition, or any other your superior officer, according to the rules and discipline of war.

Given at the office of his majesty's ordnance under our hands and seals of the said office, this twenty-seventh day of July, 1757.

J. L. LIGONIER.
CHARLES FREDERICK.
W. R. EARLE.

INSTRUCTIONS for Robert Clerk, Esq., chief engineer.

You are carefully and diligently in every respect to perform the duty of chief engineer upon the present service, and strictly to observe what is prescribed by his majesty's instructions, a copy of which, so far as relates to the duty of a chief engineer in time of action, you will receive herewith; and we require you to see that the several engineers do their duty strictly conformable to their commissions and instructions.

You are to make fair and exact draughts of all harbours and their soundings, and accurate surveys of all countries and places you come to; of attacks, trenches, intrenchments, batteries, incampments, or anything else wherein you may be employed or concerned.

You will observe that the engineers are a separate and distinct corps from the artillery, and as such the comptroller is to muster them at such times and places as shall be agreed upon betwixt you and him, and you are to order the paymaster to pay them from time to time;

and in case any of them should die or be killed on this service, you are to order the next in rank to do duty in his stead, for which end you will receive herewith a list, with the dates of their commissions; and you are to secure the effects and equipage of those so dying or killed, and to cause an inventory to be taken of the same, as soon as possible, in presence of an engineer and the clerk of the stores, so that his or their lawful representative may be satisfied of your impartiality therein; and you are to send us a duplicate thereof by the first opportunity.

You are, from time to time, to order the clerk of the stores to deliver such instruments and stores as are applicable to your branch, and he has orders not to issue any tents or other particulars appropriated for the engineers without direction in writing from you or the commanding engineer for the time being.

You are to transmit to us by every opportunity an account of any alterations that may happen among the engineers, of their conduct and behaviour, and of everything relating to the said engineers, so far as shall come to your knowledge.

LASTLY.—You are to follow all such orders and directions as shall be given you by the master-general, lieutenant-general, and principal officers of the ordnance for the time being, and to pay due obedience to the commander-in-chief of the expedition, or any other your superior officer, according to the rules and discipline of war.

Extract from the Book of General Instructions for the good government of the office of ordnance, signed by King Charles the Second, and confirmed by the succeeding kings and queens.

CHARLES R.

INSTRUCTIONS for our principal engineer.

Article 7. In time of action, or when there is intention of forming or laying a siege against any place, he is to have a draught in ground plate of the place if possible; if not, to take a careful view of its situation as near as he can, and thereof to make a draught, and to see where the attaque or attaques are most advantageously to be made; how the circumvallation and contravallation (if need be) is to be laid out and designed; and to direct and see the breaking of the ground, planting of batteries, making of platforms, conducting of trenches and mines, and to leave such engineers and conductors as will be necessary to see them carried on and executed; and to be constantly moving from one attack to another to see that all possible expedition be made; and so divide the engineers under him that they may relieve one another; and never to suffer (as far as his authority extends) any single person to be wholly intrusted with a work on an attack without he be well assured of his ability and capacity to undertake and discharge such a service.

Countersigned,

L. JENKINS.

INSTRUCTIONS for Lieutenant Richard Dudgeon, sub-engineer.

You are diligently and carefully to perform the duty of an engineer, to make fair and exact draughts of all harbours and their soundings, and accurate surveys of all countries and places you come to, of attacks, trenches, intrenchments, batteries, incampments, or anything else wherein you may be employed or concerned; and to keep a regular journal of all your proceedings, which you are to transmit to this office at your return.

You are to follow such orders and directions as shall be given you by the master-general, lieutenant-general, and principal officers of the ordnance for the time being, and to obey the commander of the forces with whom you are employed, or any other your superior officer according to the rules and discipline of war.

The like orders to:
Lieutenant Thomas Walker, sub-engineer.
Ensign Augus. Durnford ⎫
" Robert George Bruce ⎬ practitioner engineers.
William Roy ⎪
Jno. Chris. Eisen ⎭

Signed,

J. L. LIGONIER.
CHARLES FREDERICK.

INSTRUCTIONS for Mr. John Lowndes, surgeon.

You are to demand and receive into your charge and custody from the clerk of the stores the medicines and instruments provided for the sole use of the detachment of the Royal Regiment of Artillery, officers and others, and to give him your receipt for the same.

You are tô attend the private men as well as officers thereto belonging at all hours and times when called upon, and to administer proper medicines, and perform all necessary operations for their relief according to the best of your skill and judgment.

You are not to sell or dispose of any of the said medicines, or convert them to private use on any pretence whatever, otherwise you will and shall be cashiered for the same; but when the service is over you are to return the remainder of the said medicines, together with the instruments, into his majesty's stores, that you may be regularly discharged of the same.

You are to deliver a report in writing of the sick and hurt belonging to the said companies every other day, or oftener if required, to the commanding officer of artillery.

You are to follow such orders as you shall receive from time to time from the master-general, lieutenant-general, and principal officers of the ordnance for the time being, and to pay due obedience to the commanding officer of the artillery for the time being, or any other your superior officer, according to the rules and discipline of war.

Signed,

J. L. LIGONIER.
CHARLES FREDERICK.

INSTRUCTIONS for Mr. Pitman Stagg, clerk of the stores.

You are to take into your charge and care all the ordnance, powder, shot, and other stores and provisions of war issued, or to be issued, for service of the artillery embarked on board the transports now in the river Thames, and to indent for the safe keeping and regularly accompting for the same.

You are to enter into a journal or day book all the said ordnance or stores, together with your receipts and returns, expressing therein the number, nature, length, weight, scantling, quality, and other proper descriptions of the stores, from whom and whence received or returned, with the time of every such receipt or return.

You are likewise to keep a day book or journal of all stores that you deliver, distinguishing as before mentioned the quantity and quality thereof, to whom, and for what service issued, with the day of the month and date of the order authorising the same.

You are to post from the said two journals every particular species of stores into a leidger, and to keep the same constantly posted, so that at any time, on the shortest notice, you may draw out a balance of each species, and be certain of the true state thereof.

You are, at the end of any particular service for which you deliver stores, to demand an accompt of the expense thereof from the persons to whom the same were issued, which expense is to be signed by the officer in chief commanding; and you are to demand and receive back the remainder, recharging yourself therewith.

You are to give the commanding officer of artillery and the comptroller a state of your stores when and as often as either of them shall require the same.

You are not to deliver any stores (unless in cases of extreme necessity, as the hurry of actions be) without an order in writing from the commanding officer of artillery, or from the commanding engineer for the time being, for what particularly relates to the engineers; and you are always to take receipts to vouch your deliveries, without which you will not be discharged of the same.

You are to transmit to us a remain of stores by every proper opportunity, together with an account of all occurrencies relating to the artillery so far as shall come to your knowledge; and, to prevent any inconvenience that may arise from the miscarriage of letters, you are, in your second letter, to send a duplicate of your first, and of the papers therein contained; and so, from time to time, whilst you continue on this service.

LASTLY. You are to follow all such orders and directions as you shall receive from the master-general, lieutenant-general, and principal officers of the ordnance for the time being, and to pay due obedience to the commander-in-chief of the forces, the commanding officer and comptroller of artillery, and other your superior officers according to the rules and discipline of war. And you are to shew these instructions to the commanding officer and comptroller of artillery, and give each of them a copy hereof as soon as may be, together with a copy of the proportion of ordnance and stores delivered into your custody, distinguishing how the same are disposed of on board the several storeships.

Signed,

J. L. Ligonier.
Charles Frederick.
W. R. Earle.

Instructions for Mr. Pitman Stagg, paymaster.

You are to keep an exact and particular account of all money by you received and paid for service of the said artillery in such manner that it may at all times be known, if required, how much remains in your hands.

You are at the end of every month, or oftener if required, to give the comptroller of the train a state of your cash, with an account of all receipts and payments made in the said month.

You are not to pay any money for subsistence, incidents, or otherwise, without orders in writing under the hands of the commanding officer and comptroller of the said artillery, jointly or separately, as it may happen, except what relates to the engineer, for which your orders must be signed by the commanding engineer and comptroller, jointly or separately, as it may happen, expressing, if for subsistence, the sum and name of every person so to be paid, and if for incidents or otherwise, for what service or use; and you are to observe that all orders for the payments of any money, for what service soever, are to be entered in a book to be kept by the comptroller for that purpose before they are presented to you for payment, which entry is to be noted upon every order by the comptroller or such person as he shall appoint to enter the same.

You are to examine all orders for the payment of money, and to see that no more is required to be paid than what is due, and to take receipts from every person and annex them to the respective orders for vouching your payments thereof; the said receipts to be witnessed by two persons present at the payment.

If you should receive orders to pay for fascines, gabions, or such other things as are always made by the forces, and paid for out of the contingencies of the army, you must acquaint the commanding officer or comptroller they are not to be paid for by this office, and that you have no money for such uses. And this you are to do in all such like cases, taking care to give us notice that you have complied herewith, to the end that neither the commanding officer or comptroller in the hurry of business may inadvertently order payment for what may not belong to this office.

You are to enquire, in the best manner you can, into the currency of exchange with England, and, if necessary, may draw bills on us for what relates to the engineers, staff, civil officers and artificers, separate and distinctly from what relates to stores and incidents; and you may draw on James Cockburn, Esq., for the subsistence and pay of the artillery officers and men, specifying in the body of the bill, and in the letter of advice, for what particular service the same is drawn; and take care to draw at the best exchange you can, first acquainting the commanding officer or comptroller therewith, and procuring the attestation of both, or one of them, thereto, without which your draught will not be honoured.

You are to transmit to us a state of cash by every opportunity, together with an account of all occurrences relative to the artillery on this expedition so far as shall come to your knowledge, and, to prevent any inconveniences that may arise from the miscarriage of letters, you are, in your second letter, to send a duplicate of the first and of the papers therein contained, and so, from time to time, whilst you continue on this service.

LASTLY. You are to follow such orders and directions as you shall receive from the master-general, lieutenant-general, and principal officers of the ordnance for the time being, and to give due obedience to the commander-in-chief of the forces, the commanding officer, and comptroller of artillery, and all other your superior officers, according to the rules and discipline of war; and you are to shew these instructions to the said commanding officer and comptroller of artillery, and give each of them a copy hereof as soon as may be; and you must particularly remember that it is our positive *commands* that you *consult the comptroller on every occasion before any payments are made, and take his particular instructions in everything relative to your duty, either as clerk of stores or paymaster*, and strictly observe and follow his directions therein, as you will answer the contrary.

Given at the office of his majesty's ordnance under our hands and the seal of the said office, this twenty-ninth day of July, 1757, in the thirty-first year of his majesty's reign.

J. L. LIGONIER.
CHARLES FREDERICK.
W. R. EARLE.

1757. 1 August. The regiment was divided into two battalions, each having its own field officers and separate staff, and twelve companies in each, cadets included.

1 Aug. Orders by Colonel Michaelson. The king having commanded the Royal Regiment of Artillery to be divided into two battalions, his grace the Duke of Marlborough has been pleased to make the following divisions, viz.:—

	First Battalion.	*Second Battalion.*
Col.-commandant	William Belford	Borgard Michaelson.
Lieut.-colonel	Thomas Desaguliers	George Williamson.
Major	John Chalmers	Thomas Flight.
Chaplain	Montagu Barton	
Adjutant	Forbes Macbean	Duncan Drummond.
Quartermaster	Thomas Baker	John Pomroy.
Bridge-master	George Anderson	
Surgeon	James Irwin	
Surgeon's mate	William Hobson	

Companies. *Miners.*		*Companies.* *Cadets.*		
2nd	Patterson's.	1st	Ord's,	abroad.
3rd	Leith's, abroad.	4th	Skeddy's,	,,
5th	Charles Brome's, abroad.	6th	Godwin's,	,,
7th	James's.	8th	Maitland's,	,,
9th	Cleaveland's.	10th	Hislop's,	,,
11th	Hind's.	12th	Farrington's.	
13th	Joseph Brome's.	14th	Tovey's.	
15th	Hussey's.	16th	Northal's,	abroad.
17th	Gregory's.	18th	Strachey's,	,,
19th	Dover's.	20th	Martin's.	
21st	Innis's.	22nd	Smith's.	

which are for the future to be distinguished as the first and second battalions in all reports, returns, &c. The officers of the two battalions to roll in duty according to the present roster.

1757.
1 Aug. America.—Lord Loudoun's army embarked.

4 Aug. The plan of the campaign changed.

16 Aug. Their destination again changed, the army divided into parts, and sail for New York, Bay of Funday, Boston, Annapolis, Fort Edward, Fort Cumberland (Beau Sejour), and Halifax; these posts were kept in constant alarm by the French.

Orders given on board the "Ramiliss" in the Bay of Biscay.

17 Sept. "Captain James, of the artillery, is to deliver to each ship of the squadron a light brass gun of the field artillery, to be fired in the long boats at the landing of the troops, and two boxes of ammunition (half grape, half round) for them. Some of the Royal Regiment of Artillery are to work the guns."

7 Oct. This expedition returned to England unsuccessful; Captain James's company disembarked, and were quartered in Scarborough Castle.

26 Sept. Richard Glanville appointed surgeon's mate.

7 Dec. The master-general is captain of cadets.

Schemes by Lieut.-Colonel Desaguliers for improving brass and iron ordnance, and reducing their preposterous weights, by which they will prove more serviceable than the present unwieldy ones; the nation save above £500,000 in brass and iron metal over the charges of re-casting, great sums by carriage in times of marches; and our ships, that now mount only 24-pounders, be enabled to carry double the nature.

N.B.—The powder for service, salutes, and sealing, at present is about half more than should, with prudence, be suffered to be used, all which, it is hoped, will be rectified.

[1]Guns. Same weight.		Long. Ft. & In.		Guns. Same weight.		Long. Ft. & In.	
Old	New.	New.	Old.	Old.	New.	New.	Old.
24	42	8	10	4	9	6	8
18	42	8	9½	3	9	6	8
12	32	7	10	2	6	6	7
9	18	7	9½	1½	6	6	7
8	18	7	Ibid.	1	4	5	6
6	12	6	16	½	4	5	5
5½	12	6	Ibid.				

PROPORTION OF POWDER NOW ALLOWED.

Guns.	Brass. Land.	Iron. Sea.	Salute.	Sealing.
42	21 0	17 0	11 4	3 4
32	16 0	14 0	9 4	2 12
24	12 0	11 0	7 0	2 0
18	9 0	9 0	6 0	1 8
12	0	6 0	4 12	1 0
9	4 8	4 8	4 0	0 10
8	4 0	4 0	3 6	0 12
6	3 0	3 0	3 0	0 8
5½	2 10	2 10	2 8	0 8
4	2 0	2 0	2 0	0 6
3	1 8	1 8	1 8	0 4
2	1 0	1 0	1 0	0 3
1½	0 12	0 12	0 12	0 2
1	0 8	0 8	0 8	0 1
½	0 4	0 4	0 4	0 1

"I was at a proof of ordnance when two curious new invented cannon and cartridges were exhibited, tried, and approved by the Duke of M——gue, General Honywood, &c. The too affable duke referred

1 This table would appear to mean that an existing brass 24-pounder gun, 10 feet long, can be converted into a 42-pounder of 8 feet long, &c., without reducing the gross weight of metal.

them to one of the ordnance office, and desired his opinion. He replied, he did not like them. 'Your objection?' said the duke. 'I like no new-fangled inventions,' said he. Here, by the stupidity of a person who, from his post, must be deemed a judge, dropt an invention perhaps worth 100,000 such lives; too many of whom are paid large salaries or pensions, only to be idle spectators or marplot actors of their country's ruin. It is amazing to many ingenious great officers, engineers, &c., of all ages and nations, who have experienced these useful instruments of destruction, that no improvements of consequence have been made since 1335, when cannon were first invented and cast in England, and used in the battle of Cressy; and in 1535, mortars, brass cannon, &c., were first cast by Owen, at Bucksted, in Sussex.

"It is hoped this period will exert itself in giving proper encouragement (void of sinister views) to bravery and improvements that tend to national utility (either by sea or land), and which ever prove the most effective instruction. It has been cruelly hinted, I hope without foundation, that some ingenious inventions have been stifled by pretenders, who, in vain, have attempted by deviations, &c., to pass for their own what with great labour, expense, and disgrace they never could accomplish. Gunpowder, invented about 1200 by Roger Bacon's experiments, whom death did not permit to know its pernicious effects only in amusements, though it proved his most rapid messenger. Guns of 9 or 6 feet, with ball from 6 to 18, carry 400 yards point blank, and 4000 random."

Military History of Great Britain, Artillery Library.

www.ingramcontent.com/pod-product-compliance
Ingram Content Group UK Ltd.
Pitfield, Milton Keynes, MK11 3LW, UK
UKHW020112200726
13856UKWH00002B/503

9 781845 740412